China: Tackle the Challenge of Global Climate Change

Global climate change is one of the greatest challenges ever to confront humanity, with the largest scale, widest scope, and most far-reaching influence. As the biggest developing country with the largest population, China is the world's leading consumer of coal and energy, and one of the worst-hit victims of global warming. Consequently, China should assume its responsibility in making contributions to global sustainable development.

Based on the principles of fairness and efficiency, this study creatively puts forward two principles of global governance on climate change. The first entails replacement of the two-group schema of developed and developing countries with a four-group model. The second entails application of the resulting model to specify the major emitters as principal contributors to emission reduction. It also proposes a two-step strategy for China to tackle climate change. This book makes it clear that China should proactively engage in relevant international cooperation, participate in international climate negotiations, make clear commitments to reduce emissions, and assume the obligations of a responsible power to achieve sustainable and green development.

Angang Hu is one of the pioneers and leading authorities in the realm of contemporary China studies. He serves as the dean of the Institute of Contemporary China Studies and Professor of School of Public Policy and Management, Tsinghua University.

Qingyou Guan graduated from the Chinese Academy of Social Sciences with a doctorate in economics. His research interests include economic and financial theories, energy, climate change, and carbon currency.

China Perspectives Series

The *China Perspectives* series focuses on translating and publishing works by leading Chinese scholars, writing about both global topics and China-related themes. It covers Humanities & Social Sciences, Education, Media and Psychology, as well as many interdisciplinary themes.

This is the first time any of these books have been published in English for international readers. The series aims to put forward a Chinese perspective, give insights into cutting-edge academic thinking in China, and inspire researchers globally.

For more information, please visit https://www.routledge.com/series/CPH

Existing Titles

Chinese Cooperative-Harmonious Democracy
Zongchao Peng, Ben Ma, Taoxiong Liu

Scientists' Impact on Decision-making
A Case Study of the China Hi-Tech Research and Development Program
Peng Ru

China: Tackle the Challenge of Global Climate Change
Angang Hu, Qingyou Guan

Forthcoming Titles

Creative Involvement
The Evolution of China's Global Role
Yizhou Wang

China's Historical Choice in Global Governance
Yafei He

China: Tackle the Challenge of Global Climate Change

Angang Hu
Qingyou Guan

LONDON AND NEW YORK

清华大学出版社
TSINGHUA UNIVERSITY PRESS

First published 2017 by Routledge

2 Park Square, Milton Park, Abingdon, Oxfordshire OX14 4RN
52 Vanderbilt Avenue, New York, NY 10017

Routledge is an imprint of the Taylor & Francis Group, an informa business

First issued in paperback 2020

British Library Cataloguing-in-Publication Data
A catalogue record for this book is available from the British Library

Library of Congress Cataloging-in-Publication Data
A catalog record for this book has been requested

ISBN: 978-1-138-70594-4 (hbk)
ISBN: 978-0-367-52855-3 (pbk)

Typeset in Times New Roman
by Apex CoVantage, LLC

Contents

	List of figures	vi
	List of tables	vii
	Preface	ix
	Acknowledgment	xiv
1	Introduction: Climate change and human development	1
2	The international context of climate change negotiations and the dawn of the Paris Climate Conference	22
3	Global governance of climate change: Reaching a global emission reduction agreement	37
4	Climate change and China: Threats and challenges	56
5	National governance of climate change: A roadmap for China's emission reduction	78
6	Addressing climate change and achieving a low-carbon economy: The road ahead for China	95
7	Global significance and strategic consensus	126
	Postscript	153
	Bibliography	163
	Index	168

Figures

1.1   Trends in global climate change and sea level rise   5
1.2   Research framework   19
2.1   Required emission reductions for carbon dioxide equivalent (CDE) compared with the base year (1990)   25
2.2   The proportion of wealthy countries' carbon dioxide emissions in relation to global emissions (1890–2010)   30
2.3   The proportion of G8 greenhouse gas emissions to total global emissions (1992–2013)   32
3.1   The proportion of the world GDP emission reduction cost relative to the growth rate (2007–2050)   38
4.1   Proportions of heavy and light industries in China's industrialization process (2000–2012)   66
4.2   China's per capita GDP and per capita energy consumption   67
4.3   China's oil production, consumption, net imports, and foreign dependence (2003–2013)   70
4.4   China's structure of energy consumption and total amounts of energy consumed (1978–2013)   73
4.5   China's coal consumption (1990–2013)   73
6.1   The global carbon market structure (2011)   103
6.2   Number of registered CDM projects, worldwide, according to country and technology   105
6.3   Proportion of China's registered CDM projects to the total number of projects registered worldwide (as of June 2015)   111
6.4   Country-wise distribution of expected annual Certified Emission Reductions received from registered projects according to the latest forecast (2015)   112

Tables

1.1 Impacts of the rise in sea levels   6
1.2 Attitudes and interest demands of different countries or organizations in response to climate change   14
2.1 *G8 Climate Scorecards 2008*   23
2.2 Carbon footprints of OECD countries (2012)   31
3.1 Classification of global public goods   42
3.2 "One Earth, Four Worlds"   48
3.3 Top 20 CO_2 emitting countries in the world (2012)   48
4.1 The top 10 natural disasters with the highest death tolls in 2013   57
4.2 Top 10 global disasters affecting the largest numbers of people in 2013   57
4.3 The proportion of China's losses resulting from natural disasters to global losses (1979–2014)   58
4.4 Statistics on China's disaster history (Zhou Dynasty to Qing Dynasty)   59
4.5 Areas affected and damaged by floods and droughts (1950–2013)   59
4.6 Nationwide reductions in average annual grain outputs caused by natural disasters (1952–2013)   60
4.7 Nationwide economic losses directly attributed to disasters (1998–2013)   61
4.8 Distribution of arid and humid areas in China   62
4.9 China's significance within the global economy in proportion to total global volumes (1980–2013)   64
4.10 Proportion of China's main indicators to the total global volume (1950–2030)   65
4.11 International comparison of per capita primary energy consumption (oil equivalent in tons) for the period from 1820 to 2012   68
4.12 China's primary energy consumption forecast (2005–2030)   69
4.13 Proportion (%) of primary energy consumption of the top four economies to total global energy consumption for the period from 1965 to 2035   69

4.14 The proportions of CO_2 emissions of six major economies in
       relation to total global CO_2 emissions (1960–2035)   72
5.1   Two realities and two roads   79
5.2   Proportion of populations of different HDI groups to the total
       population of China (1982–2006)   81
5.3   Carbon sources, total carbon sinks, net carbon sources, and
       HDIs of various regions in China   83
5.4   Changes in regional levels of human development across
       China (1982–2010)   84
5.5   Per capita carbon emissions of China and of the world
       (1990–2013)   85
5.6   China's energy, electricity, and coal consumption growth
       and elasticity coefficients (1996–2013)   86
5.7   Recommended primary and secondary nationally prioritized
       resource and environment indicators in 2020   89
6.1   An overview of the carbon trading market (2005–2011)   102

Preface

In the 21st century, climate change has emerged as the most significant, and simultaneously the most nebulous, challenge facing human society. Its current scale, and the breadth and reach of its impacts, are unprecedented. Consequently, the forging of a global agreement on emission reduction to tackle climate change represents the most beneficial global public good (GPG) for all humankind. It is also one of the most important factors affecting the world's future economic and social development, as well as the reconstruction of the global political and economic landscape. Political leaders across the world have reached a consensus on the need to respond to global climate change. However, developing and least developed countries (LDCs) will face great inequality in emission reductions associated with the prevailing climate change mitigation trend. In recent years, political games played between countries have been evident, prompted by the unevenness of costs borne and revenues shared by them. An apparent prisoner's dilemma has constrained cooperation, making climate change governance problematic. Because the Kyoto Protocol is a minimally binding international convention, it lacks not only necessary punitive measures, but also the most basic incentives. Several disputes have arisen because developed countries, which are the source of most greenhouse gas emissions, are not willing to take initiatives to fulfill the emission reduction targets, insisting that developing countries currently are the bigger greenhouse gas emitters. There are several ongoing disputes over stocks and the flows, the right to survive and develop, and the right to consume luxuries. These have created an "emission reduction paradox" within global climate change governance that was directly responsible for the failure of the 2009 Copenhagen Conference.

At the climate conference held in Paris on December 12, 2015, 195 contracting states of the United Nations Framework Convention on Climate Change (UNFCCC) adopted the first global agreement that comprehensively addressed emission reduction, adaptation, loss, funding, transparency, capacity building, and other key elements. This agreement signaled a turning point in the history of negotiations on global responses to climate change. In particular, significant progress has been made in relation to previously stalled core issues pertaining to the overall objective of global climate change governance, the division of responsibilities, capital, technologies, and other salient matters that emerged after the Copenhagen Conference. Consequently, the Paris Agreement has been acknowledged as a historic agreement, reflecting the "equal responsibilities and obligations; coordinated strength, ambition and development; matched actions and supports; balanced emission reduction and adaptation; and convergence of strength before 2020 and actions after 2020." Moreover, it demonstrates important characteristics, such as comprehensiveness, balance, intensity, and legal impacts. Although no reduction in emissions, or a low reduction, would be the best option from the perspective of short-term economic interests, this would cause suffering in all countries, regardless of whether they were developed or developing, in the long term. Thus, while the far-reaching significance of the Paris Agreement will only become fully apparent by 2030 or 2050, it nevertheless marks a major historical turning point in the global response to climate change. From a historical perspective, the future of humankind is the outcome of our own choices, and these choices are not only diverse, but also unpredictable.

China's dual roles relating to the global response to climate change have given rise to some key considerations. The first consideration relates to China's status as the world's largest consumer of coal and energy, as well as the largest emitter of sulfur dioxide and carbon dioxide. China is also among the countries most affected by global climate change. Consequently, it is China's responsibility to respond unflinchingly to this issue. However, as a developing country, China is unable to assume the same level of responsibility that developed countries can assume. From the perspective of historical responsibility, climate change can largely be attributed to carbon emissions released by Western developed countries during their industrialization process that evolved over a period of several hundred years. Therefore, developed countries should assume greater responsibility in this regard.

Second, having the largest population worldwide and a fragile natural environment, as well as previous experience of grave natural disasters, China is one of the countries most affected by global climate change. China must therefore respond with a positive attitude and join in efforts to advance global governance and develop international agreements to tackle climate change. However, current levels of awareness regarding the environment and climate security are low within the country. The contradiction between economic development and improvements in the environment and climate remains prominent, and the Chinese public still hesitates on the issue of emission reduction when facing the prospect of economic losses.

Third, as a responsible nation, China has the obligation and capacity to promote energy conservation and respond to climate change. On November 30, 2015, President Xi Jinping gave a keynote speech entitled "Work Together to Build a Win-Win, Equitable and Balanced Governance Mechanism on Climate Change" during the opening ceremony of the Paris Climate Conference. He emphasized China's active engagement in the global campaign on climate change and its intention to work hard to implement the vision of innovative, coordinated, green, open, and inclusive development, fostering a new pattern of modernization that favors harmony between humans and nature. At the same time, he pointed out that China ranks highest in the world in terms of energy conservation and utilization of new and renewable forms of energy. He noted that the proposed goals for China's "intended nationally determined contributions" will require strenuous efforts for their achievement. However, the Chinese government has the confidence and the resolve to fulfill its commitments. With the establishment of the China South-South Climate Cooperation Fund in September 2015, China has pledged to earnestly fulfill its policy commitments of South-South cooperation in relation to climate change. It will continue to promote international cooperation in areas such as clean energy, disaster prevention and reduction, ecological protection, climate-adaptive agriculture, and low-carbon smart city construction, while also actively helping other developing countries to enhance their financing capacities. However, it is noteworthy that in the area of public opinion, there is a wide gap between China and the developed countries, especially in relation to energy-saving technologies and energy efficiency. Consequently, the country still has a long way to go in advancing its emission reduction policies.

The final consideration is that climate change is a universal problem facing all of humanity. Thus, laxity and refusal to cooperate on the part of any one country may engender global failure. The world's major carbon emitters, including China, are particularly important in this context, as their inability or reluctance to make common cause with others would undermine international cooperation and lead to its failure to respond effectively to climate change.

Currently, climate change has evolved from being a global environmental issue to being a "carbon politics" issue. China's strategy is essentially to pursue its interests under the established framework of rules. In contrast to other international negotiations, China has engaged in carbon politics from the onset, gaining familiarity with and participating in the development of its rules. Nonetheless, to date, China has not formulated a unified strategy for addressing carbon politics that encompasses its political, economic, social, and cultural aspects.

Based on a scientifically grounded understanding of climate change and of the challenges relating to its global governance, this study proposes a global governance and emission reduction scheme to address climate change. In accordance with this scheme, it outlines a roadmap for China's accomplishment of emission reduction.

The study has been conducted according to the principles of fairness and efficiency. More than 200 countries have been classified into four groups, as opposed to two groups based on the traditional binary division. Appropriate emission reduction contributions were determined based on the proportion of the world's major emitters to the total amount, globally. We creatively posited two principles of global governance. The first entailed replacement of the two-group schema of developed and developing countries with a four-group model based on the Human Development Index (HDI). The second entailed application of the resulting model to specify the major emitters as principal contributors to emission reduction. Our study clearly revealed that there are 49 countries in total, accounting for approximately 17% of the world's entire population (nearly 1.2 billion people), which belong to the extremely high HDI group. These countries should play a principal part in emission reduction as unconditional emission reducers. They should, therefore, be obliged to follow the relevant emission reduction principles of the United Nations (UN). According to the data on HDI trends (1980–2013) contained in the UNDP's *Human Development Report 2014*, there were 40 countries and regions in the extremely high HDI group in 2005. As an increasing number of countries enter the extremely high HDI group, the number of countries and regions required to unconditionally fulfill their emission reduction obligations will correspondingly increase. Currently, there are 11 countries in the extremely high HDI group that are categorized as unconditional mandatory emission reducers and 6 countries in the high HDI group, categorized as conditional emission reducers. Only India, Indonesia, and South Africa belong to the medium HDI group, although India, which ranks as the third-greatest emitter in the world, should actively implement emission reduction. When India enters the high HDI group, its status will correspondingly change to that of a conditional but active emission reducer.

Political leaders of all countries agree on the need to respond to climate change. The focus of the current global debate is not about whether climate change exists; rather it is on how to respond to climate change. Thus, China's agenda is not whether to participate in international climate negotiations, but how to secure the right to speak and a position as one of the new global rule-makers. Although many people still do not realize the significance of China's response to climate change, an article entitled "The New Sputnik," published in the *New York Times* on September 27, 2009, articulated the dramatic impact that China will have. It noted: "The most important thing to happen in the last 18 months was that Red China decided to become Green China." The article also pointed out that "China's leaders, mostly engineers, wasted little time debating global warming. They know the Tibetan glaciers that feed their major rivers are melting. But they also know that even if climate change were a hoax, the demand for clean, renewable power is going to soar as we add an estimated 2.5 billion people to the planet by 2050, many of whom will want to live high-energy lifestyles. In that world, E.T. – or energy technology – will be as big as I.T., and China intends to be a big E.T. player." Moreover, the article provided the following warning to the United States and Western countries: "no doubt . . . China is embarking on a new, parallel path of clean power deployment and innovation. It is the Sputnik [the first satellite launched by the Soviets] of our day. We ignore it at our peril." Thus, the manner in which China responds to climate change and to "carbon politics" will evidently determine its economic development, with proactive efforts serving to enhance its political status, internationally, and even jump-start future long-term national revival.

From the perspective of global economic development trends, efforts to address climate change and to reach a global emission reduction agreement will be accompanied by a new industrial revolution – the fourth in humanity's history – the Green Industrial Revolution. Although China did not have the opportunity to be part of the first and second industrial revolutions, and was a follower during the third one (the Information Revolution), the country is now poised at the same starting line as the United States, the European Union, and Japan to become a leader, innovator, and driver of this fourth revolution. To assume this position, China needs to actively participate in global emission reduction actions, reaching global emission reduction agreements together with other countries, especially the great powers. By making environmental contributions to human development, China will give tangible form to the popular expression "one world, one dream, one action."

Acknowledgment

On the completion of this book, we would like to express our deepest gratitude to all my colleagues and students, whose hard work and efforts have made this work possible. Doctor Yan Fei has been directly involved with the writing, organization of material, research and calculations, editing, and modification, for which he deserves respect.

1 Introduction

Climate change and human development

Climate change is a global issue, and its impacts on the development of societies are causing worldwide concern. In addition to bringing about ecological, economic, and social disasters, global climate change has contributed to inequitable income distribution and even to national security challenges. Consequently, the issue of how to mitigate these effects constitutes a global challenge. The Paris Agreement, adopted in 2015, entails specific modalities for implementing concerted action as a global response to the climate change challenge post-2020. This agreement has enabled the international community to finally emerge from the shadows of the fruitless Copenhagen Conference.

1.1 The research context of climate change: The considerable challenges it poses for human development

Ecologists have long been concerned about resource depletion and environmental pollution problems.[1] From the 20th century onward, a spate of books and papers has been published on the subject. In the early 1970s, these issues were first presented by the Club of Rome as a "global issue." By the 1990s, the question of how humankind should deal with global climate change had become a topic of interest within academic research circles, as well as an important policy concern for governments.

Climate change refers to the emergence of statistically significant change in the average state of the climate and/or a deviation (an anomaly) from it. An increase in this deviation indicates an increase in climatic instability.[2] Climate change, as defined by the Intergovernmental Panel on Climate Change (IPCC), refers to any change in climate over time, including changes caused by natural factors as well as human activities. Climate change is defined in paragraph one of the United Nations Framework Convention on Climate Change (UNFCCC), which specifically refers to climate change caused directly or indirectly by human activities.

Global climate change is itself a highly controversial topic. There are still a number of unresolved arguments about climate change within the natural sciences. For example, it is forecast that the global surface temperature will continue to rise by 1.1 °C to 6.4 °C over the next 100 years. This estimated figure is the outcome of 58 value tests conducted by various countries, applying 23 climate system models. The question that arises is whether these mathematical models accurately reflect natural variations and the scope of human activities. There are many uncertainties in this regard.[3] Some research institutions believe that global climate change is merely the result of Earth's own warming and cooling transitional cycle, which has nothing to do with human activities. There has always been climate change throughout human history. The climate has warmed and then cooled time and again, and the determination of warming or cooling is also related to which base year and temporal scale are selected. If we conduct an evaluation of climate change, applying the temporal scale of the last century, it is an indisputable fact that global climate change has been markedly characterized by warming. Because of long-term changes in the climate system and its inherent complexity, there is evidently still a long way to go before we can clearly explain the issue of climate change.

Data analysis[4] has revealed fluctuations in the earth's temperature over the past 100 years, with an overall rise of 0.5–0.6 °C. The average surface temperature rise of the Northern Hemisphere has been in the range of 0.6–1.4 °C, with the temperature rise corresponding to an increase in latitude. Thus, the temperature has risen by about 0.4 °C at latitudes of 0°–30°N, by about 1 °C at latitudes of 30°–60°N, and by about 2 °C at latitudes of 60°–90°N.

Ultimately, the question remains as to whether the earth's environment has gradually become warmer because of human activities. Will this warming trend eventually result in unparalleled disaster for humans? Global climate change is not only a scientific issue, but also a policy issue. Although a considerable number of scientists have warned of the effects of climate change, there are also individuals within the scientific community who have challenged the theory of global warming. One viewpoint is that a periodic change occurs in the earth's climate itself. Earth experienced the Little Ice Age during the 17th and 18th centuries. Thus, according to this viewpoint, Earth's temperature rise from the late 19th century has only signaled the end of this ice age. Proponents of this view also point out that global warming is caused by many factors, including solar activity and changes in cosmic rays. Some scholars whose perspectives are more extreme even believe that humans do not have the ability to produce significant changes in nature, and, therefore, climate change caused by human activities is nothing but an argument reflecting a sense of self-importance. This divergence in views is reflected in the terminology used. Those who believe that global warming is caused by human activities refer to it as a "global warming phenomenon," while skeptics are more likely to refer to it as "climate change." "Global warming" and "climate change" are combined into one term as an eclectic argument.

General understanding of climate change has undergone a shift from a conception of climate change as a static and stable phenomenon to one that is dynamic and abrupt. Before the 1970s, it was believed that climatic characteristics could be described based on an assessment of the climate over an average period of 30 years. The concept of a climate system was proposed in the 1970s, based on the recognition of climate changes occurring at various temporal scales. However, in the 1980s, a new approach was proposed for understanding the earth's system based on holistic and dynamic changes of the planet. After 1990, people began to recognize the abrupt character of climate change, and the changing temporal scale has decreased from 1,000 years to 10 years. There are over 100 reasons that explain climate change, and natural factors, including changes in Earth's orbit, solar radiation, and volcanic eruptions, among other reasons, may contribute to this phenomenon. Currently, human activities, in addition to natural factors, have aggravated the impacts of global climate change. In the context of natural climate cooling, global warming over the past century is especially likely to exhibit an unnatural trend because of the significant impact of human activities. The main reasons for global warming are widely attributed to the greenhouse effect, caused by excessive emissions of carbon dioxide (CO_2) and other greenhouse gases (GHGs) into the atmosphere.[5]

On the whole, and over the entire history of humankind, climate change has undergone numerous alternations between cold and warm periods. However, the rates of such extensive natural changes were much slower before humans. Human activities over the past century have fundamentally differed from natural factors that have historically contributed to climate change. An aspect of climate change that is of grave concern is whether it is periodic or whether the cumulative effects of GHGs will lead to a continuous monotonic temperature rise. Climate change that occurs within such a condensed temporal scale presents a huge challenge for human adaptation. It is also in this sense that climate change, as an environmental issue, enters the purview of the international community.

The IPCC released four global climate assessment reports in 1990, 1995, 2001, and 2007, which constitute the main scientific foundation for the international community's awareness and understanding of climate change issues. In 2014, the IPCC published its fifth climate change assessment report. This report presents the IPCC's scientific conclusions based on its compilation of collaborative research involving the world's leading scientists. It represents authoritative and state-of-the-art knowledge of human society in relation to climate change.[6] By now, all countries have generally accepted that the concept of global climate change originates in the natural sciences. The IPCC's Fifth Assessment Report clearly reveals that human activities are the main cause of global climate change. The CO_2, methane (CH_4), and dinitrogen monoxide (N_2O) concentrations in the global atmosphere have increased significantly from 1750 onward because of the impact of human activities, and they now far exceed pre-industrialization levels based on polar ice core records for periods dating back thousands of years. The polar core records reveal an increase in the concentration of CO_2 from about 0.280 ml/L prior to industrialization to 0.395 ml/L in 2012. The concentration of CH_4 has increased from about 7.15×10^{-4} ml/L before industrialization to 1.859×10^{-3} ml/L in 2012, and the concentration of N_2O increased from about 2.70×10^{-4} ml/L before industrialization to 3.24×10^{-4} ml/L in 2012. The impact of human activities has been manifested as an overall warming of the climate, with a radiation intensity of $+ 2.29$ W·m^{-2}.

Currently we are observing rising average air and sea temperatures, widespread melting of snow and ice, and a rising average sea level, globally. The global climate is presented with significant changes characterized by warming. Over the past three decades, from 1983 to 2012, the average linear warming rate (0.28 °C per year) was almost four times that of the rate over the last 100 years, and the total temperature rise from 1983 to 2012 was 0.85 °C. The rate of the temperature rise in the middle and lower layers of the troposphere is similar to that recorded for surface temperature. The main temperature change can be attributed to the global temperature rise of oceans, as the temperature 75 m under the surface of the sea rose, on average, by 0.11 °C per annum during the period of 1971 to 2010. Mountain glaciers and accumulated snow in the Northern and Southern Hemispheres have retreated on the whole. During the period of 1901 to 2010, the average rate of the global mean sea level rise was 1.7 mm per annum, and the rate during the period of 1993 to 2010 increased by about 3.2 mm per annum. The sea level rise throughout the entire 20th century was estimated to be 19 cm (see Figure 1.1).

The IPCC's fifth report, released in 2014, also pointed out that most of the observed rises in the average global temperature during the last 50 years were likely to have been caused by increased anthropogenic GHGs (at a probability of 95%). Currently, evidence of recognizable human activities has extended to other aspects of the climate, including ocean temperature, water cycle changes, ice and snow melting, and temporal changes in manifestations of extreme climate. An analysis of previously unexplored simulated results under observation constraints provides the possible range of climate sensitivity and an understanding of the response of the climate system to radioactive forcing. In a scenario entailing the doubling of CO_2 concentrations, the average magnitude of global warming may be 2–4.5 °C higher than it was prior to industrialization. The most accurate estimate is about 3 °C, and the magnitude of warming is unlikely to be less than 1.5 °C. The average global surface temperature for the period of 2016 to 2035 is expected to rise by 0.3–0.7 °C, compared with that during the period of 1986 to 2005, and by 0.3–4.8 °C by the end of the century. Moreover, the global sea level will continue to rise at a rate that exceeds the annual rate of 2.0 mm during the period of 1971 to 2010. These estimates reveal that by 2030, global CO_2 emissions will increase by 45%–100% from their level in 2000. Two-thirds to three-fourths of this growth in CO_2 emissions from energy sources will primarily be from developing countries. By the end of this century, the sea level will have risen by 60 cm.

Global climate change is evidently creating significant challenges for human development. The first of these challenges relates to ecological disasters. The IPCC's Fifth Assessment Report has stated that within the next 10 years, 1.1 billion people worldwide will face a shortage of drinking water. By the middle of the 21st century, a further 130 million people in Asia will be threatened by starvation. By 2100, crop revenues in Africa will be reduced by 90%. By 2100, the rising sea level will cause the greatest loss of agricultural land in Bangladesh compared with that experienced by any other country in the world. By 2050, rice and wheat production in Bangladesh will be reduced by 10% and 30%, respectively.

(a) **Globally averaged combined land and ocean surface temperature anomaly**

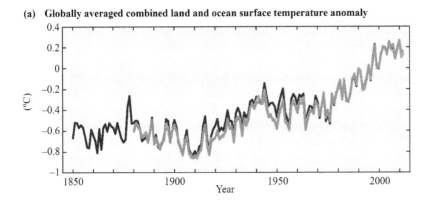

(b) **Globally averaged sea level change**

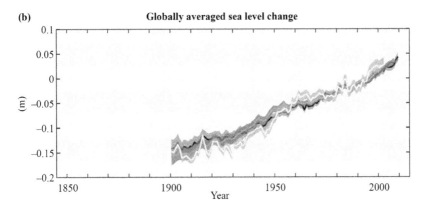

(c) **Globally averaged greenhouse gas concentrations**

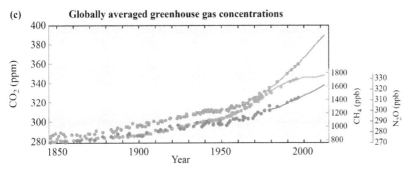

Figure 1.1 Trends in global climate change and sea level rise
Source: The IPCC's Fifth Assessment Report (2014), 3.

According to the estimates of some scientists, global climate change will lead to a historically unprecedented scale and pace of human migration, involving 200 million people worldwide. In its *Human Development Report 2007/2008: Fighting Climate Change: Human Solidarity in A Divided World,* the United Nations

Development Programme (UNDP) has confirmed that climate change is a scientifically proven fact. It is difficult to predict the precise impact of greenhouse gases, and there are many uncertainties relating to scientific predictive capability. However, according to current studies, there is sufficient information to confirm that huge disaster-related risks exist. These are likely to be catastrophic. They include the melting of ice sheets in Greenland and the western Antarctic region (resulting in the possibility of many countries being submerged) and diversion of the Gulf Stream (which may bring about dramatic climate change). A rise in sea levels will have a considerable impact on economic and social development, globally. According to projections, a rise in the sea level of 1 m (the probability of this occurring is 70%–80%) will affect 0.3% of the land area, 1.3% of the population, 1.3% of the GDP, 1.0% of the urban area, 0.4% of the agricultural area, and 1.9% of the wetland area across the globe. Small island states such as the Maldives, a country that is only 1.5 m above sea level, are likely to be submerged. If current global warming trends continue, the Maldives will disappear during this century. On October 17, 2009, the former Maldivian president, Mohamed Nasheed, held the world's first underwater cabinet meeting on the country's seabed at a depth of 4 m. A resolution titled "SOS from the Frontline" was signed at this meeting and was subsequently submitted at the United Nations Climate Change Conference held in Copenhagen that same year, making a strong plea for countries to reduce their GHG emissions. In November 2013, Nasheed announced that to prevent the continued rise in the sea level and to guard against the loss of homes, the Maldives would set aside a portion of its tourism income for the purchase of new land to arrange new accommodation for the country's 386,000 inhabitants. Clearly, we cannot turn a deaf ear to appeals from countries affected by the ecological disaster caused by climate change.

The second challenge is economic disaster. A report overseen and completed by Nicholas Stern, a former chief economist for the World Bank, provided the first assessment ever made of the economic impact of global warming in US dollars. The report simultaneously presented a picture of the future of human society. It revealed that if we do not take timely action over the next few decades, the

Table 1.1 Impacts of the rise in sea levels

Sea level rise (m)	Impact (percentages of total global amounts)					
	Land area	Population	GDP	Urban area	Agricultural area	Wetland area
1	0.3	1.3	1.3	1.0	0.4	1.9
2	0.5	2.0	2.1	1.6	0.7	3.0
3	0.7	3.0	3.2	2.5	1.1	4.3
4	1.0	4.2	4.7	3.5	1.6	6.0
5	1.2	5.6	6.1	4.7	2.1	7.3

Source: United Nations Development Programme, *Human Development Report 2007/2008: Fighting Climate Change: Human Solidarity in a Divided World*, 2007.

economic and social crises resulting from global warming will be comparable to those brought about by the two world wars and the Great Depression during the first half of the 20th century. By then, the loss in the global GDP will have reached 5%–20%.[7] According to an estimate published in *Nature* in 2015, the global per capita GDP will fall by 23% within 85 years unless we intervene in relation to GHG emissions.

The third challenge confronting us is social disaster and inequitable distribution. The IPCC's report contends that some people will face the threat of death as a result of global warming. Because of their limited technical and financial capacities for responding to disasters, the poorest countries will be most severely affected. Heat will cause the spread of dengue fever, cholera, malaria, and other infectious diseases in Africa, resulting in increased mortalities across the continent. The abilities of countries, worldwide, to adapt to climate change are highly unequal. The UNDP report notes the following:

> People living in both the Ganges Delta and Lower Manhattan face the risk of flooding because of [a] sea level rise, but they have various vulnerabilities: there is a high level of poverty in [the] Ganges Delta, and a lower level of infrastructure protection. When tropical storms and flood[s] occur in Manila, Philippines, the whole city is at risk. However, the vulnerability focuses on the crowded slums on both sides of [the] Passy River, rather than the more affluent areas of the city. Currently, rich countries are led by the[ir] government[s] to invest heavily in climate change defense systems. By contrast, developing countries are neglected and marginalized, which may lead to "apartheid" of the adaptation to climate change. Long-term ecological challenges posed by climate change are not primarily concentrated in Manhattan or London, but in flood-prone Bangladesh and Africa to the south of the arid Sahara.

The last challenge relates to national security. Climate change not only threatens economic development, but also has serious impacts on national security. An EU report[8] notes that climate change will worsen existing trends, tensions, and instability. In particular, the following seven threats emanate from climate change:

- Resource-based conflicts
- Threats to coastal cities and infrastructure
- Loss of territory and border disputes
- Environmentally induced migration
- A gradual societal decline and increasingly widespread aggressive behaviors
- Shortage in energy supplies
- Increasing international regulatory pressures

It is clear that if we fail to take timely action, the costs of national defense and security, as well as the costs induced by conflict related to climate change that are incurred by countries worldwide, will far exceed the costs of reducing emissions. According to estimates by the UN, most of its emergency humanitarian assistance efforts in 2014 were related to climate change.

In recent years, studies on climate change and national security issues have shown that climate change is becoming a major threat to national security. In October 2003, the United States Department of Defense submitted a confidential report to the Bush Administration titled *An Abrupt Climate Change Scenario and its Implications for United States National Security*. This subsequently aroused worldwide concern. In April 2007, the Military Advisory Board of the Center for Naval Analysis in the United States published an article titled *National Security and Climate Change Threat*. This article assessed the potential threat posed by climate change to the national security of the United States from a military perspective. Further, in November 2007, the United States Center for Strategic and International Studies and the Center for a New American Security launched a report titled *The Age of Consequences: The Foreign Policy and National Security Implications of Global Climate Change* that envisaged the impacts, in different contexts, of climate change on international and American security. It summarized the top security threats faced by the United States as follows. The "soft power" issue will become increasingly prominent; tensions between the North and South will increase; domestic and cross-border migration will increase, with grave consequences; public health problems will become increasingly serious; resource conflicts and vulnerability will increase; nuclear activities and risks will increase; global governance will face more severe challenges; domestic political instability and instances of state failure will be evident; and military changes will occur in unpredictable ways. China's role will be vital, and it will be essential for the United States to actively respond to climate change.[9] In November 2007, the US Council on Foreign Relations made specific recommendations to the government of the United States in its report titled *Climate Change and National Security: A Program of Action*. The report noted that while climate change does not pose a threat to the basic survival of the United States, it nonetheless poses a direct threat to the country's national security. In June 2008, on behalf of 16 national intelligence agencies in the United States, the National Intelligence Council estimated the possible impacts of future global climate change on the security of the United States. These estimates were submitted to the government in a confidential report titled *National Intelligence Assessment on the National Security Implications of Global Climate Change to 2030*. The concluding section of the report has now been made public. The report contended that the regions that would be most seriously affected by climate change were sub-Saharan Africa, the Middle East, Central Asia, and Southeast Asia. Spillover effects of climate change, especially migration and water resource disputes, are likely to have adverse global impacts. However, unlike *The Age of Consequences,* this report does not conclude that climate change will lead to "state failure." Moreover, large-scale wars between states over scarce water resources are not inevitable. In the fall of 2008, the journal *Security Studies* published a paper by Joshua W. Busby, a scholar at the University of Texas, titled "Who Cares about the Weather? Climate Change and U.S. National Security." This paper represented an attempt by the academic community in the United States to present theoretical research on the impacts of climate change on national security. Busby analyzed the effects of climate change on the security of American interests abroad according to four categories of threats: strategic threats, moral challenges, threats that require monitoring, and weak threats. In doing so, he considered four dimensions, namely American property abroad, violent conflicts, failed states, and humanitarian catastrophes.[10] In January 2008, the Oxford Research Group, based in the United Kingdom, released Chris Abbott's report titled *An Uncertain Future: Law Enforcement, National Security Climate Change*. The report noted that over the next 10 years, climate change was set to generate new domestic and international threats to the security of the United Kingdom. The report stated that rising temperatures would lead to rising sea levels and changes in weather patterns, and the world would face wars resulting from food shortages and a lack of clean water. The international community would face the pressure resulting from the displacement of hundreds of millions of "environmental refugees." These refugees would have to flee their homes because of disasters. For the United Kingdom, pressure on the security sector would increase as a result of the implementation of stricter border controls, violent conflicts arising from public protests against polluting companies, and the diffusion of tensions within communities. The report further noted that governmental measures to mitigate climate warming would likely be resisted, which also posed a major obstacle to the development of new laws limiting GHG emissions. Chris Abbott has also warned that attempts to apply existing policies to the governance of new problems will eventually fail. Government leaders must therefore develop new ways of cooperating to mitigate and manage climate change.[11] In March 2011, the United Kingdom launched nationwide climate change response options. In the United States, the Central Intelligence Agency (CIA) began funding research projects on climate change in 2013. Further, in November 2013, China first issued its "National Strategy on Adaptation to Climate Change" to safeguard its national economic security. More recently, in May 2015, President Obama called for action, stating that climate change threatened the national security of the United States.

Thus, the impacts of climate change on human development are wide ranging and pressing, requiring effective measures to tackle them. The IPCC's Fifth Assessment Report notes that global warming caused by climate change, and especially by GHG emissions, is having a significant impact on the global society and economy. Urgent measures are therefore required to stabilize future GHG emissions and to avoid serious negative impacts on the global environment, society, and economy. However, there is an economic cost to be paid for achieving this. Currently, the concentration level of global GHGs amount to 396 ppm (parts per million) of carbon dioxide equivalent. Let us suppose that it is possible to control the GHG concentration peak value at 445–710 ppm by 2030. The maximum global GDP loss would consequently be 3.2%. However, unless existing climate change emission reduction policies and measures are improved, global GHG emissions will continue to increase in the coming decades.

1.2   Outline of the issues: The context of global emission reduction and the deadlock in climate talks

Given the grim situation regarding global climate change, reaching a global emission reduction agreement is an urgent requirement. In 1992, the UN Environment and Development Conference adopted the UNFCCC, which became the first international legal document to address climate change. It came into force on March 12, 1994.

The Kyoto Protocol, adopted in December 1997, became the first enforceable document requiring signatories of the treaty to meet their obligations for protecting the earth's climate system. The Protocol was creative in its formulation of a Clean Development Mechanism (CDM) and emission trading system. China signed the Kyoto Protocol in May 1998. In November 2005, the decision to adopt the Kyoto Protocol was taken at the UN Climate Change Conference held in Montreal, Canada. The Kyoto Protocol subsequently entered the full implementation phase.

In 2007, the IPCC published its Fourth Assessment Report on climate change. During the same year, the International Energy Agency proposed the "450 Stabilisation Case," aimed at stabilizing the atmospheric CO_2 equivalent concentration at around 450 ppm and limiting the average global temperature rise to 2.4 °C above that which existed before the industrial revolution. Further, in December 2007, the "Bali Roadmap" was proposed at the UN Climate Change Conference.

In May 2008, Nicholas Stern presented his "Key Points to Address Climate Change," aimed at controlling the future concentration of the atmospheric GHG CO_2 at 450–500 ppm. Global GHG emissions will peak in 2020, and these emissions will need to be reduced by 50% by 2050 compared with levels in 1990, with a per capita emission amount of two tons. Subsequently, on July 8–9, 2009, a decision was taken at the G8+5 Summit to reduce GHG emissions by 80% by 2050, and to limit the temperature rise to no higher than 2 °C above pre-industrial levels. This was followed by the Copenhagen Summit, convened by the UN in December 2009, where a global emission reduction agreement was decided to take place in 2012.

In 2010, the UN climate change negotiations were held in Bonn, Germany. The work plan for these negotiations was adopted at the conference, and a consensus was reached on the basis for future negotiations and the holding of two further meetings before the Cancun Conference that was tabled for the end of 2010.

From April 3 to 8, 2011, the first round of UN climate change negotiations for that year was held in Bangkok, Thailand. A consensus was reached on the negotiating agenda for the conference that year. During the first round of UN climate change negotiations held in 2012 from May 14 to 25, in Bonn, Germany, a consensus was reached on establishing an "Ad Hoc Working Group on the Durban Platform for Enhanced Action" to pursue the agenda laid out at the climate conference held in Durban the previous year. Moreover, relevant legal and technical details relating to the second commitment period of the Kyoto Protocol were clarified and progress was also made regarding long-term financial support and other matters.

At the 2012 Doha Conference, the Kyoto Protocol was extended to 2020. However, Japan, Canada, Russia, and New Zealand announced their withdrawal from the second compliance period (2013–2020) of the Kyoto Protocol. This was followed in 2014 by Australia's legislative repeal of the carbon emission tax and of the emissions trading scheme, the implementation of which was planned for 2015.

At the 2013 Warsaw Conference, participants reaffirmed the importance of the implementation of the Bali Roadmap and reached a decision to further promote the Durban platform. The following year, in 2014, the IPCC's Fifth Assessment Report emphasized that to achieve the targeted temperature rise of less than 2 °C, global GHG emission levels would need to revert, at the least, to their levels in 2010 by 2030. They would further need to be reduced by 40%–70% of their 2010 levels by 2050, approaching near-zero emission by 2100.

At the UN Climate Summit held in September 2014, participants committed to the universal climate deal that was eventually reached at the Paris Climate Conference. The public and private sectors identified financial channels, enabling the implementation of a carbon pricing mechanism through a variety of means.

At the Lima Climate Summit held in December 2014, participants agreed on the basic elements of the draft agreement for the 2015 Paris Conference and defined the submission rules of the "intended nationally determined contributions" (INDCs).

In June 2015, the first round of the 2015 UN climate change negotiations was held in Bonn, Germany. Consultations on the new climate change agreement continued, and by the end of the year, a decision was made to step up efforts to respond to climate change, and other issues, by 2020.

At the Paris Climate Conference held on December 12, 2015, 195 contracting parties of the UNFCCC reached the Paris Agreement, the first global agreement on climate change to be signed in Paris. Although clearly not perfect, the agreement clarified the programs and directions for the future global response to climate change, representing a historic turning point for global climate negotiations.

For some time now, international organizations and national politicians have made concerted efforts to explore and find a common emission reduction target and program that would be acceptable to more than 200 countries and regions worldwide. Evidently, they are all pursuing their own interests, and an ongoing debate centers on whether to reduce emissions based on the two core principles of the UNFCCC. These are the "common but differentiated responsibilities principle," and the obligation of all countries, and especially developing countries, to implement sustainable development. If the answer to this question is affirmative, then a further question arises as to what proportion of emission reduction should be shared to form a cooperative and united body of developed and developing countries, as opposed to creating an incentive mechanism for more "free-riders." It remains difficult to reach a global emission reduction agreement because of the prevailing interests and demands of respective countries. A global emission reduction agreement cannot be reached in the event of the failure of the United States, the EU, and Japan, as well as of China, India, Russia, and other countries and regions to reach a consensus. The 2009 Copenhagen Summit, considered a failure, has already been consigned to history, but learning from this experience remains a worthwhile endeavor. Experience has demonstrated that it is unrealistic to expect countries to voluntarily address the climate problem. The Paris Agreement reached at the 2015 UN Climate Change Conference signals a dawn of hope and the momentum to make further progress.

Nevertheless, we must acknowledge the possibility that in recent years, the global governance response to climate change has steadily evolved into a political game played between countries, causing new initiatives in international cooperation to revert to the circumstances of a prisoner's dilemma. The main reason for the impasse in climate talks lies in the different positions held by the parties involved that are associated with three major inherent contradictions in international climate negotiations, outlined below.

(1) The contradiction between developed and developing countries. Developing countries argue that climate change is caused by the large quantities of GHGs emitted during the industrialization process implemented by developed countries. Consequently, the loss caused by climate change should be borne by these countries. Developed countries, in turn, argue that it is difficult to assess the impact of human-induced climate change using current technology. Moreover, it is difficult to assess the costs and benefits of climate change adaptation. Moreover, the current understanding of the effects of taking adaptive measures to reduce the impacts of climate change is poor.

(2) Diverse viewpoints among developing countries. Least developed countries (LDCs) and the Alliance of Small Island States (AOSIS) have stressed the need for assistance in implementing concrete adaptation measures. Oil-exporting countries stress the economic impacts caused by emission reduction measures taken by developed countries, and emphasize the need for support for economic diversification. Relatively large developing countries emphasize the importance of technology transfers and capacity building for adaptation to climate change.[12]

(3) Contradictions also exist between developed countries, mainly between Europe and America. Currently, the EU is expected to fulfill the emission reduction obligations specified in the Kyoto Protocol for the following primary reasons.

  • Global warming is having a significant impact on Europe. The weakening or disappearance of the North Atlantic Drift would have a major impact on the whole of Europe. Major European countries are thus highly motivated regarding the issue of emission reduction out of a sense of urgency.

  • The energy consumption of major European countries has shifted from coal and oil to natural gas, demonstrating a significant increase in energy efficiency.

  • Following the eastward expansion of the EU, new Eastern European member countries have widened the space for overall emission reductions within the EU as a result of the economic downturn.

- The EU hopes to strengthen its "soft power" via the climate change issue, expanding its political influence and seizing the commanding heights of the moral watchdog and maker of international rules.

Located in the zone of the mid-latitudes, the United States is less affected by global climate change compared with countries in other zones. This reality coincides with the obstruction of efforts to address climate change posed by domestic interest groups, an example being the initial refusal of the Bush administration to ratify the Kyoto Protocol. By contrast, the Obama administration has taken a markedly more positive attitude toward climate change. In August 2015, the Obama administration issued the most stringent "Clean Power Plan" in history, which entailed a reduction of GHG emissions from power stations in the United States by one-third over the next 15 years. Moreover, by 2030, carbon emissions from American power plants would be reduced by 32% compared with the 2005 levels. This plan, however, met with resistance within the country. In October of the same year, 24 states jointly filed a lawsuit against President Obama's clean power policy, accusing Obama of exceeding the government's statutory authority in calling for a reduction in the use of traditional fossil energy sources and promoting vigorous development of wind and solar energy. On February 11, 2016, the Supreme Court of the United States voted against and ultimately rejected the Obama administration's Clean Power Plan. This was a significant blow to the efforts of the United States to promote clean energy.

The contradictions between the Global North and South constitute the most significant issue during international climate negotiations. Developing countries adhere to the "common but differentiated responsibilities principle," failing to assume the specified emission reduction obligations, and thereby causing discontent among some of the other countries. An important excuse offered by the United States for refusing to ratify the Kyoto Protocol is that major emitters like China and India have not made an emission reduction commitment. The EU has also stressed that large developing countries like China should meet their emission reduction obligations. Emission reduction has become an important issue featuring in China's relations with the United States and the EU. In recent years, although China has attached growing importance to the construction of an ecological civilization, it remains unrealistic for it to completely abandon economic development to reduce its carbon emissions. Russia is likely to benefit from growth in agricultural productivity. It did ratify the Kyoto Protocol, but only because by doing so it could actually obtain the right to emit GHGs, which could prove highly lucrative.[13] Because of the lack of consensus on GHG emissions among developed countries and between developed and developing countries, the international community is faced with a prisoner's dilemma in relation to cooperation on climate change issues, resulting in the ongoing deadlock in climate change negotiations (Table 1.2).

Global climate change is the biggest constraint, challenge, and context affecting China's future economic and social development. The economic rise of China has changed global patterns, resulting not only in unprecedented challenges to China's own energy security and environmental protection, but

also unprecedented global challenges relating to energy and climate change.
From 1978 to 2014, China's GDP grew twenty-eight-fold at an average annual
growth rate of 9.7%. China's total industrial output increased forty-five-fold, and
the average annual growth rate of industry reached 11.2%. Compared with other
large and rising countries, China broke the world record for economic growth
and continues to show rapid expansion. From a global perspective of modern
economic development history, China is experiencing a period of rapid rise and

Table 1.2 Attitudes and interest demands of different countries or organizations in response
to climate change

Country or organization	Basic attitudes	Interest demands
EU, United Kingdom, and Australia	Positive	Continue to strictly limit greenhouse gas emissions in industrialized countries and include developing countries in the institutional framework for emission reduction; adopt flexible actions to reduce emissions, combined with advanced emission reduction technologies; cooperate under the framework of the United Nations.
United States	Relatively negative, but has demonstrated an improved attitude in recent years	Repeal all rigid emission reduction targets; reduce pollution through the use of new environmental technologies under a free market model; strive to realize economic development and environmental protection; insist that large developing countries make greater efforts.
Least developed countries, economic transitional countries, and Russia	Positive	These countries have lower greenhouse gas emissions and do not face any pressure to reduce their emissions. Consequently, they advocate comprehensive emission reduction.
Developing countries undergoing rapid development	Relatively positive	Greenhouse gas emissions are increasing every year, and there is a strong demand for emission-releasing development. These countries are facing the most pressure to implement environmental protection. However, they continue to emphasize the route of voluntary commitment, and insist on practical considerations of countries' rights to use resources and fair distribution.
Energy exporters and Japan	Negative	Emission reduction will cause a tightening of global energy markets and damage their economies. Japan's economy is weak and there is no room for further energy saving.

Source: Prepared by the author based on the progress of international climate negotiations.

of strategic opportunity. China offers the world its creation and demonstration of a vast new development opportunity rather than constituting a deadly threat. In short, the rise of China has not only profoundly transformed the country, but will also reshape the world. China entered the economic takeoff stage after embarking on its reform and opening up in 1978. According to purchasing power parity (based on the international dollar rate in 1990), China's GDP accounted for just 4.5% of the world's total GDP in 1978. The International Monetary Fund (IMF) predicted that by 2015, China's GDP would account for 15.5% of the total global GDP, indicating an increase of 4% compared with the amount in 2012. At the same time, the gap between China's economic aggregate and that of the United States had significantly narrowed by 2015, with China's GDP being equivalent to 63.4% of that of the United States, an increase of nearly 11% compared with the rates in 2012. After 1978, China truly entered an era of "favorable climatic, geographical and human conditions," which is historically extremely rare. It embarked on a period of great rejuvenation, resulting in its relative prominence and marked by its significant contributions to human development.[14]

While enjoying the developmental fruits of China's economic rise, the world must also face the negative effects associated with the rise of such a large developing country. **During this rise, China's demand for and consumption of resources, particularly energy, has become a global focus of attention.**[15] The *2007 World Energy Outlook: China and India Insights,* released by the International Energy Agency (IEA), provides comprehensive insights into the "legitimate aspiration" of China and India to improve the living standards of 2.3 billion people. With expanding affluence in these two countries, more energy will be consumed within office buildings and factories, and the number of purchased home appliances and cars will increase. These developments greatly improve the quality of life of their populations. This evidently also requires tolerance and support extended by other countries around the world based on the recognition that such a strong desire is reasonable.[16] However, the rapid growth of China's economy is also undeniably bringing about unprecedented challenges that are reflected in the following two prevailing trends:

- An increasing scale of energy demands and growing dependence on foreign sources
- Excessive demands for fossil fuels, leading to the deterioration of the ecological environment and accelerating climate change

China's rapid economic development over the past three decades has come at a high cost to the ecological environment, generating all kinds of pollution, which in turn has caused both economic losses and the loss of natural assets. Large quantities of emitted carbon are causing air pollution, and are even resulting in an acid rain phenomenon in some parts of China. The quality of urban air in China is far below international standards, and at least 270 million urban residents live in an environment affected by very serious air pollution. China is now the second-largest emitter of CO_2, and is also the largest emitter of other major pollutants, including CH_4, N_2O, dust, black carbon, and sulfur dioxide (SO_2). Moreover, the proportions of these other emissions exceed those of CO_2 in relation to global quantities by varying degrees. The effects of these emissions are being endured by the Chinese people.

At present, China is undergoing a critical period of reform transformation and economic slowdown. On the one hand, China's economic rise has brought benefits to the world economy, and China has become a new global player. Thus, the world economy is sharing the fruits of China's economic development. On the other hand, issues associated with China's rise, including those relating to energy, the environment, and climate change, undoubtedly pose a great number of challenges to the world at large. To tackle such challenges requires the involvement of all countries, including China, in the creation of a new governance model for realizing rational use and enhancement of global energy and the environment.

Currently, China needs to engage with global trends in responding to climate change. Despite China's increasing emphasis on the construction of an ecological civilization in recent years, Chinese scholars have continued to insist on the "common but differentiated responsibilities" principle in relation to China's position as a developing country. To some extent, this stance hinders the ability of decision makers to acquire a correct understanding of how to deal with climate change in a timely and accurate manner, and how to participate in international climate negotiations. China is evidently a large carbon emitter, and there is considerable disparity between China and Europe and the United States in terms of energy cleanliness. This study suggests that China should make mid- and long-term emission reduction commitments to the international community, bearing defined emission reduction obligations. China should set phased, step-by-step, and industry-specific emission reduction targets. Based on a comprehensive assessment of the potential of domestic industries to achieve emission reduction, these targets should be considered as the bottom line for China during climate negotiations. If joining the World Trade Organization (WTO) is perceived as China's participation in international competitions, then a clear commitment regarding emission reduction obligations will be perceived as a move toward cooperation with developed countries. China's clarification of its emission reduction obligations presents an open and credible commitment to the international community. This study aims to elaborate on this idea further, providing arguments in its favor, conducting a feasibility analysis, and also making an assessment on China's meeting its emission reduction commitments.

There are two major aspects for China to consider in dealing with the challenges posed by global climate change. The first entails the costs and benefits of emission reduction and climate change investment. The second concerns China's strategies and responses relating to its participation in international climate change negotiations. These two aspects are mutually interactive. China should proactively engage in international cooperation on climate change, actively participate in international climate negotiations, make clear commitments to reduce emissions, and assume the obligations of a responsible power to achieve sustainable and green development of its economy and society. Doing so will promote the establishment of a more harmonious world and allow China to ascend to a leadership role in dealing with global climate change. This study argues that challenges can be transformed into opportunities, and China can make environmental contributions to the world through its own green development and its rise as a "green power." China's success is the success of the world, and China's opportunities correspondingly entail opportunities for the world. Humanity must contend with global issues relating to energy, the environment, and the climate, and China's challenges are simultaneously global challenges. Solving and dealing with these problems will determine the future of humanity. Consequently, all countries need to act in unison to transform challenges into opportunities.

Responding to climate change requires worldwide cooperation as opposed to confrontation, and requires working together to deal with issues and create opportunities for common development. In this process, China, as one of the permanent members of the UN Security Council, should assume the responsibility of a large country in making contributions to sustainable development on a global scale.

As a latecomer to industrialization, China has the opportunity to avoid past mistakes made by other countries and to create a new development model centering on a low-carbon economy. An appropriate way of responding to global climate change is not to stop development, but to make fundamental changes in the means of development, pivoting away from "black" toward "green" development, from excessive exploitation of nature toward harmonious coexistence, and from "ecological deficit" toward "ecological surplus." In essence, this requires the implementation of the concept of scientific development. China's response to climate change must succeed to avoid global failure, and our fight for China's success is simultaneously a fight for achieving the success of the entire world in tackling climate change.

1.3 Analytical framework: Political economy for responding to climate change

This study adopts the concepts of international political economy to examine climate change issues. Thus, climate change is viewed as a specific research area within the field of international political economy. International political economy is itself a branch of international political science that specializes in studies of the relationship between international politics and international economics.[17] Its research methods are characterized by a strong orientation toward issues, and its research field tends to focus on a specific issue or area such as the international trade issue, the international monetary system, the global commons, global public goods, international organizations, international rules, and international environmental issues. Some of these issues are interrelated, and of considerable theoretical and practical significance.

Climate change is a typical "global issue." It is not only a global environmental issue, but also a global "tragedy of the commons," caused by excessive use of global public environmental resources by countries worldwide.[18] Global issues have a relatively strong impact on all nations and demonstrate very strong negative externalities. The solving of global issues (through a governance model or program) actually serves to provide global public goods (GPGs) that have strong positive externalities. Any country, acting alone, faces difficulties and is reluctant to provide GPGs. There are three routes for providing these goods: through hegemonic countries, international organizations, and joint provision by multiple countries. Because of the inherent deficiencies relating to the provision of public goods, we must design an "incentive-compatible" international mechanism for encouraging the provision of public goods. Addressing climate change depends on the provision of GPGs comprising a global climate program or governance. Responses to global climate change and fostering good governance based on global climate policy through international climate negotiations are science-based economic and political events that, although interconnected, entail huge policy differences. "Governance" refers to the process of achieving public management by means of the government and market. In accordance with the definition agreed upon by the Organization for Economic Cooperation and Development (OECD), "good governance" means that a government body ensures that regulatory agencies run efficiently and democratically to secure the common interests of the nation and the world through the implementation of reasonable, coherent, and extensive procedures.[19]

Currently, the consensus of countries worldwide is that governments must take steps to curb as well as adapt to global climate change trends. Climate change is not merely a scientific issue but also an economic and political one. Most countries recognize that climate change constitutes a common challenge for humanity and needs to be dealt with through individual efforts combined with international cooperation. However, because of inequalities relating to cost bearing and the sharing of proceeds, international climate negotiations and cooperation are not operating smoothly.

The climate change issue entails a two-level game.[20] Because a government is subjected to pressure on both domestic and international fronts, its external climate policy often demonstrates fluctuation around its core objectives. This fluctuation depends on the game being played among all of the domestic interest groups as well as on the pressure being exerted on the government. For example, the United States Congress has variously adopted three positions on climate change. The first is one of complete opposition to taking measures to address climate change. The second is an environmentalist position demanding legislation for climate change. The third is a mercantilist position advocating a response to climate change by the United States while also considering economic benefits. There are particular interest groups behind the different cliques within the Congress that hold these positions and, inevitably, the attempts to balance the views and interests of these various groups are reflected in American policy in relation to international climate negotiations. At the international level, other countries regard the United States as the largest emitter of GHGs. The Bush administration came under considerable international pressure when it refused to ratify the Kyoto Protocol, and the Obama administration has begun to appeal for reductions in carbon emissions at a time when its term is coming to an end, but it is facing considerable obstruction within the country. Consequently, acquiring the complete trust of other countries is proving difficult.

This study mainly assesses the climate governance issue at the global and national levels. At the global level, addressing climate change requires a global governance model and reduction program in which all countries participate. At the national level of China, the government needs to ensure the consistency of its international and domestic policies, and to participate in a global emission reduction agreement to address climate change. We propose a global and national governance framework for this purpose. Reaching a global emission reduction agreement and implementing actions to reduce emissions are at the core of global governance. At China's national level of governance, taking the initiative to publicly commit to emission reduction, specifying the emission reduction time table up to 2050, and announcing a national emission reduction plan are at the core. China must develop practical, effective, and feasible emission reduction targets and new energy goals. The conceptual framework of this study is presented in Figure 1.2.

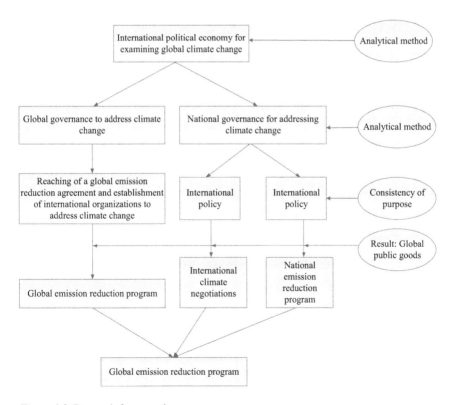

Figure 1.2 Research framework

Notes

1 Tom Titans Berg, *Environmental and Natural Resource Economics* (Beijing: Economic Science Press, 2003), 3.
2 The Coordination Group Office on Climate Change, the Management Center for Agendum in the 21st Century, *The Global Climate Change: Challenges Faced by Humanity* (Beijing: Commercial Press, 2004), 17.
3 See the speech of Zheng Guoguang, administrator of the China Meteorological Administration and head of the Chinese delegation that participated in the Governmental Review Meeting conducted for the fourth assessment of the Intergovernmental Panel on Climate Change. Zheng Guoguang, *"Correctly Understand and Deal with Global Warming," China Environment Observer* 1 (2007).
4 Ruan Junshi, *Ten Lessons on Meteorological Disasters* (Beijing: Meteorological Press, 2000), 172.
5 The greenhouse effect means that while greenhouse gases allow visible sunlight to pass through the atmosphere to reach the earth's surface, they also absorb most of the infrared rays radiated by the earth, resulting in temperature rises. See Zhang Qiang, Han Yongxiang, and Song Lianchun, "Overview of Advances in Research on Global Climate Change and Its Effect Factors," *Advance in Earth Sciences* 9 (2005): 990–998.
6 For the full report, see the official website of the Intergovernmental Panel on Climate Change: www.ipcc.ch/.
7 Nicholas Stern, "Stern Review on the Economics of Climate Change," *NBER Working Paper 12741.* www.nber.org/papers/w12741.pdf.
8 The High Representative and the European Commission to the European Council, "Climate Change and International Security." www.consilium.europa.eu/ueDocs/cms_Data/docs/pressData/en/reports/99387.pdf.
9 Kurt M. Campbell, Jay Gulledge, J.R. McNeill, et al., "The Age of Consequences: The Foreign Policy and National Security Implications of Global Climate Change," *Center for Strategic and International Studies.* www.sallan.org/pdf-docs/071105_ageofconsequences.pdf.
10 Joshua W. Busby, "Who Cares about the Weather? Climate Change and U.S. National Security," *Security Studies* 17 (2008): 468–504.
11 Chris Abbott, "An Uncertain Future: Law Enforcement, National Security and Climate Change," *Oxford Research Group.* http://oxfordresearchgroup.org.uk/sites/default/files/uncertainfuture.pdf.
12 See Li Yu'e and Li Gao, "Status of Negotiations on the Climate Change Impacts and Adaptation," *Advances in Climate Change Research* 3 (5) (2007): 304.
13 Eric Posner and Cass Sunstein, "Pay China to Cut Emissions," *The British Financial Times* (Chinese edition), August 9, 2007.
14 Hu Angang, *China in 2020: Building Up a Well-Off Society* (Beijing: Tsinghua University Press, 2007), 104.
15 China's rise has changed the future profile of the world's energy structure and its energy use in three ways. First, in relation to the international energy market, growing demands for energy within large emerging market economies have intensified the contradiction between supply and demand, while also increasing market expectations of rising energy prices to a certain extent. However, this has also broadened market prospects for the majority of energy-exporting countries. Second, in relation to the political pattern associated with global energy, as countries with very high demands, the status and roles of China and India are becoming increasingly strong. Their interests are being increasingly aligned with those of most energy-consuming countries. Consequently, they are becoming key stakeholders representing these countries. This has increased the overall right of energy-consuming countries to speak, as well as their negotiating ability. Energy-consuming countries have the choice of cooperation or confrontation. Cooperation between them means that they rarely stand or fall alone. Third, the

prevailing Western high-energy consumption pattern and the current OECD-dominant global energy governance pattern are unable to adapt to the rapid increase in China's and India's energy demands. Moreover, they also increase the insecurity of the world's economic growth for energy security, prompting changes in the energy consumption patterns of Western countries, as well as changes in the existing economic development and global energy governance models.

16 International Energy Agency, Executive Summary of *2007 World Energy Outlook: China and India Insights* (in Chinese).

17 It has also been suggested that international political economy is primarily a research method that focuses on international political issues. In short, it represents an attempt to penetrate and analyze international politics and relations from an economic perspective. See Zhang Yuyan and Li Zenggang, *International Economic and Political Studies* (Shanghai: People's Publishing House, 2008), 9.

18 This term is derived from the work of an American scholar, Garrett Hardin, entitled *Tragedy of the Commons,* which was published in 1968 in the journal *Science*. According to Hardin, the United Kingdom once had a land system wherein feudal lords set aside an area of uncultivated land within their own territory as pasture (the "commons"), which was open to herders at no cost. This land was originally intended to benefit the people, but as it was free grazing land, every herder tried to raise as many cattle and sheep as possible. This uncontrolled increase in the number of cattle and sheep and the resulting "overload" ultimately led to the common pasture becoming barren land. Consequently, the herders' cattle and sheep eventually starved to death. The application of the "tragedy of the commons" within economics refers to welfare losses associated with an inefficient state and caused by the abuse of public resources.

19 Organization for Economic Cooperation and Development, *Governance in China (Chinese Version)*, trans. The Institute for Contemporary China Studies Tsinghua University. (Beijing: Tsinghua University Press, 2007), 308.

20 Putnam (1988) proposed a two-level game model for analyzing international politics that has been used to analyze the interaction between international and domestic politics. In his view, international and domestic politics are inextricably linked, and rulers must consider the levels of both domestic and international politics in the formulation of foreign policy. At the level of domestic politics, various interest groups try to influence the policy formulation process, so that the policies that are finally introduced are in their best interests. Ruling parties consider the situation and make all efforts to maximize their own legitimacy through alliances forged with various interest groups. At the international level, negotiators sent by government heads, or even the heads themselves, are seeking to meet the requirements of their interest groups during the negotiations, and to enable their countries to maximize their gains, at the expense of other countries. See Robert D. Putnam, "Diplomacy and Domestic Politics: The Logic of Two-Level Games," *International Organization* 42 (1988): 427–460.

2 The international context of climate change negotiations and the dawn of the Paris Climate Conference

Developed as well as developing countries are required to participate and cooperate in global governance initiatives responding to climate change. Evidently, progress toward the emission reduction targets specified in the Kyoto Protocol has been unsatisfactory. Consequently, the development of emission reduction options that are acceptable to all countries is necessary to reach broad consensus regarding global emission agreements. It is also necessary to dismantle the traditional divide between developed and developing countries in relation to emission reduction obligations. As the largest emitter of carbon dioxide, China can and should actively reduce its emissions in response to climate change. The financial crisis of 2008 constrained international negotiations on global climate change, which resulted in the fruitlessness of the Copenhagen Summit. The slow recovery of the global economy following the crisis has also, to some extent, inhibited country-based implementation of energy-saving/emission-reducing actions. Currently, despite the evident recovery of the developed economies, represented by the United States, the outlook for the world economy as a whole is not optimistic, particularly following the slowdown within emerging market economies, especially China. Some urgent structural problems that have accumulated over many years have now come to light. Resolving all of these to achieve sustainable economic growth will take more time, thereby posing new challenges for global climate change initiatives. However, it must be emphasized that the economic impacts of the financial crisis are short-term ones, whereas climate change constitutes a long-term threat to humanity.

2.1 Considerable efforts required to meet emission reduction targets

None of the developed countries, represented by prominent Western countries, have succeeded in achieving the emission reduction targets specified in the Kyoto Protocol. According to the report produced by Ecofys, entitled *G-8 Climate Scorecards 2008,* not one of the developed and industrialized countries, categorized as the "Group of Eight" (G8) nations, fulfilled their emission reduction targets (see Table 2.1)[1] aimed at preventing a rise in the global temperature above 2 °C. Moreover, the United States performed worst among these nations. According to

Table 2.1 G8 Climate Scorecards 2008

Rank	Country	Overall performance indicators
1	United Kingdom	The United Kingdom is expected to reach its Kyoto target and to introduce innovative policies such as a climate change bill. Although the United Kingdom maintains a focus on the carbon market, it has made minimal investments to accelerate the use of renewable energy and promote energy efficiency. In addition, the United Kingdom's share of the coal energy structure is on the rise, which also increases emissions.
2	France	France ranks second in terms of its current objectives, performance, and international status. However, France will find it difficult to achieve its objectives in the future. Consequently, its relatively high ranking cannot be guaranteed in the next scorecard.
3	Germany	Germany demonstrates the best performance in relation to renewable energy, and its regulatory framework is in compliance with international benchmarks. Germany has also passed new legislation on energy efficiency and enacted a climate policy. However, so far, it has not taken a clear stance against coal power generation. Instead, it demonstrates an increasingly negative trend in relation to its electricity consumption and its plan to establish a large number of new coal and lignite power plants.
4	Italy	Italy has begun to make some efforts to address climate change issues and has endorsed the EU's policy to derive associated benefits. However, Italy can only implement a few specific measures, and its emissions greatly exceed its Kyoto target. However, the country has a relatively good ranking in terms of energy efficiency.
5	Japan	Japan's emissions continue to increase, extending ever further from its Kyoto targets. Nevertheless, the government has yet to announce a midterm emission reduction target. Japan ranks the second because of its wide application of carbon offset projects in developing markets, based on the Kyoto Protocol's Clean Development Mechanism. However, it lacks mandatory measures at the national level, such as emission trading.
6	Russia	Russia has lost its early advantage of low emissions. Over the past eight years, its emissions have increased. Only a few national policies have been formulated, and these remain to be implemented. The Russian government has recently announced that it will significantly increase energy efficiency. This may have a positive impact on Russia's scorecard ranking in 2009.
7	Canada	With rising emissions and an energy-intensive economy, Canada has failed to realize the potential of promoting energy efficiency.
8	United States	The performance of the United States has been the worst, evidencing the greatest quantities of total and per capita emissions and a rising trend relating to emissions. The United States will shortly be enacting legislation for restricting emissions. Moreover, enterprises are getting ready for a new carbon emissions trading market that has the potential to cover the entire continent. While the federal government, over two terms, did not support a climate-friendly program, various states have taken initiatives that may help improve the country's ranking in 2009.

Sources:
1. Website of the Allianz Insurance Group (www.allianz.com.cn).
2. Ecofys, G8 Climate Scorecards: Performances of Canada, France, Germany, Italy, Japan, Russia, the United Kingdom, and the United States of America. Background information on China, Brazil, India, Mexico, and South Africa, www.panda.org/about_wwf/what_we_do/climate_change/publications/g8_nations.cfm?uNewsID=138001.

the British newspaper *The Guardian*, carbon emissions of the world's Fortune 500 companies increased by 3.1% in 2013, compared with emissions in 2010, far exceeding the emission standards for greenhouse gases (GHGs) developed by the United Nations.[2] Currently, it is difficult for the G8 countries to achieve the emission reduction targets specified in the Kyoto Protocol. Thus, for the "Group of Five" (G5), comprising the five developing countries invited to participate in the G8 Summit, to insist on the requirement that developed countries reduce their emissions in accordance with the targets specified in the UNFCCC and the Kyoto Protocol lacks pertinence. Moreover, the emission reduction targets specified in the Kyoto Protocol cover broad regions. This will not encourage these countries to reduce their emissions.

As signatories of the Paris Agreement, developed and developing countries have succeeded in temporarily avoiding the prevailing prisoner's dilemma relating to global governance of climate change. However, the actualization of commitments and the effectiveness of actions taken will have to stand the test of time. Following the G8 Summit held in July 2008, the participating G5 emerging countries issued statements pointing out that developed countries should take the lead in emission reduction after 2012, reducing their emissions by at least 25%–40% and 80%–95% by 2020 and 2050, respectively, based on emission quantities in 1990. However, without the active and simultaneous participation of developing countries in this effort, particularly China, India, and other large countries, achieving the global emission reduction target will be difficult, and the situation may even worsen. It is indicated that if no action is taken, GHG emissions will exceed 60 billion tons by 2030, and will rise to over 85 billion tons by 2050 (Stern, 2008).[3] Thus, as the time it takes to launch emission reduction increases, the amount of emission reduction required also grows, entailing increasingly radical measures for achieving the desired outcome (see Figure 2.1). Therefore, developing practical, feasible, and effective emission reduction targets and timetables are top priorities for all countries. Developing countries account for more than 75% of the world's population. Consequently, although these countries face the same impacts of climate change, the damage that they endure as a result is much greater than that endured by developed countries. The G5 countries, as the fast-growing economies, including China and India, should face up to this reality and its serious consequences. These countries should represent the long-term interests of developing countries, change their previous stance of not fulfilling quantitative emission reduction targets, and actively participate in global emission reduction. By doing so, they may actually succeed in breaking the impasse prevailing within global climate change negotiations and governance.

According to the latest report by Stern and his colleagues, entitled *Key Elements of a Global Deal on Climate Change*,[4] the majority of future emissions will emanate from developing countries, given their rapid population and GDP growth and given that they account for an increasing proportion of energy-intensive industries. More aggressive emission reduction targets are needed to reduce overall emissions worldwide to 500 ppm or less. On the one hand, it will take a long time for the impact of our current actions on future climate change to manifest, and

(Unit: 1 billion)

Global GFG emissions, billion tonnes CO_2e

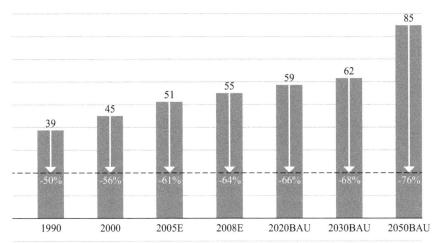

BAU = business as usual

Figure 2.1 Required emission reductions for carbon dioxide equivalent (CDE) compared with the base year (1990)

Source: Tony Blair, "Breaking the Climate Deadlock: A Global Deal for Our Low-Carbon Future." *Report submitted to the G8 Hokkaido Toyako Summit (2008)*. http://blair.3cdn.net/b53ed18e b4812ef5d3_dem6be45a.pdf.

this will be limited during the next 40 to 50 years. On the other hand, our actions over the course of the next 10 to 20 years will have a profound impact on the climate during the second half of this century and the next century.[5] To ensure the reduction of global emissions by 50% before 2050, developed countries will need to reduce their emissions by at least 80% by 2050, while China's emissions need to be reduced by 50% before 2050. The situation is very clear. Unless China can reduce emissions by 50% before 2050, achievement of the overall global emission reduction target of 50% before 2050 will be difficult. For China to achieve tenfold growth of its economy by 2050, compared with its size in 2005, as well as reduction of its GHG emissions by 50%, compared with emissions in 2005, it will have to reduce emissions per unit by one-twentieth of the value in 2005 by 2050. This means that by 2050, China's emission per unit would have to be reduced to 5% of the current value (an emission reduction of 95% per unit).[6]

Some economists do not accept that limiting emissions is an effective strategy for tackling climate change. One of the earliest studies on climate change was conducted by the Nobel laureate Thomas Schelling, an economist who also served on the Carbon Dioxide Assessment Committee of the National Academy of Sciences from 1981 to 1983. In Schelling's view it is not possible for the international community to formulate a global treaty that specifies GHG atmospheric concentrations, because there are no universal guidelines on limits of GHG concentrations that have been borne out by experience over the last 25 years. It is also impossible for the international community to know what emission limits should be defined for each country. Moreover, even if such limits could be specified, no countries could achieve them, and no sanctions can be imposed on countries unless these are compliant with the Kyoto mechanism.[7]

2.2   Developing an emission reduction program that is acceptable to all countries

On June 27, 2008, a report written by the former British Prime Minister Tony Blair et al., entitled *Breaking the Climate Deadlock: A Global Deal for Our Low-Carbon Future* (hereinafter referred to as the Blair Report), was released in Tokyo, Japan. The report was also presented to G8 Leaders, with the political objective of providing a decision-making context for the G8 Summit that was held in July 2008 in Hokkaido, Japan.[8] As a former statesman, Tony Blair has in-depth knowledge and experience of the failure of simple persuasion to resolve political differences or conflicts of interest.

Convincing evidence is required to persuade political leaders to reach a consensus and cooperate so as to resolve these issues. Consequently, the Blair Report includes a comprehensive description of the choice to be faced by leaders in the next 18 months, and it highlights some important considerations related to climate-related negotiations, as outlined below.

First, the report clearly notes that the world is expecting that the most advanced countries will take the lead on this issue. The statistics presented by the authors reveal that between 1980 and 2005, the emissions of the powerful "G8+5" nations (i.e., 13 countries in total) accounted for 68%–71% of the world's total emissions. During the last decade, because of annual increases in developing countries' carbon emissions, this figure has dropped to around 60%. Evidently, without the cooperation of major emitters, the climate change deadlock will continue. Moreover, the existing global governance framework does not provide a way to resolve the climate issue. The Blair Report expresses concerns regarding current global institutional capacities for addressing climate change. It points out that although the UNFCCC and the Kyoto Protocol have provided a sound institutional base, they do not themselves entirely fulfill the requirements for effectively managing responses and actions needed to address climate change.

Second, the Blair Report informed political leaders that while economic losses resulting from efforts to address climate change have little impact on national economies, delayed action will increase emission reduction costs. Referring to a number of previous studies, the report reveals that the average cost of reducing global emissions by 2050 would entail a reduction of the annual GDP growth rate by just 0.12%. This reduction in the growth rate is negligible, and will have less impact on the economy than will interest rate adjustment, inflation, or fluctuations in business cycles. The report reiterates that this conclusion should serve to allay the worries of political leaders that emission reduction will deal a disastrous blow to their national economies. Instead it constitutes the basis for settling this issue. However, the report warns that if the year of commencing emission reduction is postponed from 2010 to 2020, the annual global emission reduction will double. Therefore, it urges the world to act now.

The Blair Report further recognizes that the current political dilemma is not a "whether or not" question; it is a "how" one. This is because countries that initially displayed a passive attitude in their response to climate change have realized that climate change is "our" problem and not just "yours." To put it another way, climate change is a shared problem that affects us all and no one can evade it. All of us are victims in relation to the impacts of global climate change. Consequently, both developed and developing countries need to assume their respective responsibilities. There is no doubt, however, that developed countries should bear the primary responsibility. The Blair Report points out that the United States has agreed that developed countries should bear primary responsibility for the recent cuts in GHG emissions. Under the leadership of Prime Minister Yasuo Fukuda, Japan's perspective on this topic changed in a positive way. In Europe, there is a genuine and deeply felt consensus on the need to act. The report expresses the hope that developing countries will join the ranks of nations committed to reducing their emissions. To avoid extreme climatic risks, all countries should adjust their economic structures and reduce carbon dioxide emissions; commitments made by developed countries are insufficient on their own.

Last, the Blair Report provides a number of potential reference solutions. It expresses high hopes for the G8 leadership. The first step in the proposed program is, therefore, to develop long- and medium-term targets for global emission reductions under the leadership of the G8 Summit. The Blair Report also expresses high hopes for developing countries and notes the need for emerging market economies to increase carbon productivity during the next few decades. It further proposes a system and operational mechanisms, and professes full confidence in the role of the global carbon-trading market. A strong global carbon-trading market, it suggests, will be able to reduce the costs of emission reduction by 50%. However, historical experience shows that projected costs are often higher than actual costs, as it is difficult to predict carbon prices or the responses to technological innovations or other incentive measures.

The Blair Report explores 10 crucial elements within the Copenhagen agenda for establishing an effective, economical, and equitable convention. In this regard, it recommends that industrialized countries should assume a leading role, commit to working with developing countries, and reduce absolute carbon dioxide emissions exceeding the amount specified in the Kyoto Protocol. Fulfillment of these commitments would provide the key to developing a carbon-trading mechanism, supported by national action plans. Many countries would engage in domestic carbon trading, with all of these countries simultaneously participating in global carbon trading. Developing countries would also need to contribute by slowing down their growth and reducing emissions. These contributions and their time frames should reflect different proportions of emissions and capacities among nations.

The Blair Report contends that one of the most significant challenges, in relation to both the Copenhagen Summit and the implementation of a comprehensive international climate policy framework, will be to strengthen the UN Framework Convention on Climate Change (UNFCCC). The institutional arrangements for implementing the policies related to this treaty and for delivering on its objects will need to be designed and agreed on. An institutional framework dominated by the UN and Bretton Woods is unlikely to be able to meet this challenge.

The Blair Report raised two very important issues:

(1) The initiative to achieve a global emission reduction agreement needs to start with the powerful countries, namely the G8 nations.
(2) Developed countries must be redefined and developing countries classified so as to clarify their obligations and enable the latter to move into the ranks of nations responsible for emission reduction.

Blair made a special visit to China to meet with Premier Wen Jiabao. During this visit, he stated the abovementioned views in person. In fact, China has been subject to external pressures emanating from developed countries. In response, China reiterated that it was a developing country with differentiated responsibilities. However, this position did not convince the developed countries; nor could it detract from China's dilemma as the world's major emitter. The above two issues, raised in the Blair Report, are in fact related to the issue of how to classify countries involved in the global emission reduction agreement, which is crucial for achieving the success of the agreement. Currently, in addition to the emission reduction program entailed in the Kyoto Protocol, many scholars have independently proposed climate policies and emission reduction programs.

In their introduction, Jing, Jing, and Li (2008) noted that the environmental economics project of the Kennedy School of Government at Harvard University proposed six programs. Of these programs, two – the "graduation and deepening" mechanism, and a dynamic map of countries participating in emission reduction – were related to a nation's development stage. Axel Michaelowa, a German climate policy expert, proposed the graduation and deepening mechanism. He recommended the development of quantitative and legally binding country-specific targets for achieving a long-term and stable global carbon dioxide emission target of 550 ppm. Specifically, this enables the inclusion of all countries in the policy framework through the application of a graduation mechanism. Countries with emission commitments can still conduct international trade in carbon emissions, while developing countries can also gradually begin to accept emission targets through the use of a graduation mechanism. In other words, when the per capita emissions and GDPs of developing countries exceed the critical value, they should bear some responsibility for excessive energy use and emission of carbon dioxide. If a country arrives at a higher threshold through graduation over time, its emission targets will be tightened so as to match its emissions and revenue at that point in time. Before surpassing this quantitative target set for their graduation thresholds, developing countries have the opportunity to domestically implement their Business as Usual (BAU) Clean Development Mechanism (CDM) projects, and to develop policy frameworks that promote a shift in CDM from being a project-based approach to being a policy-based mechanism.

David Victor, a professor of law at Stanford University, has designed a dynamic map for depicting the emission reductions of participating countries. This program is very similar to the program of the L20 (a forum for leaders of 20 major industrialized and developing countries) proposed by former Canadian prime minister Paul Martin. Victor's proposal of this dynamic map was originally intended to facilitate the establishment of an agreement framework among a few countries that were key players in climate change, and to carry out negotiations within a limited venue such as the L20 forum. The participating countries would consequently commit to a number of domestic climate policies and measures.[9]

The fundamental objective of the graduation and deepening mechanism is to gradually undertake emission reduction obligations according to the dynamic development of countries. However, this program does not specify standards relating to emission reduction obligations. While the dynamic map of countries participating in emission reduction identifies the countries that should undertake emission reduction obligations according to their emission stocks, it only considers emission stocks. It does not consider emission flows, per capita emissions, and total emissions, or historical and current responsibilities.

There are two core principles underlying the UNFCCC. These are the principles of common but differentiated responsibilities, and the obligation on the part of all countries, particularly developing countries, to promote sustainable development. Global emission reduction obligations follow a shared principle, and there are different policies for developed countries. However, in the real world, while both classification methods are very broad, definitions of developed countries, such as the 27 OECD member states, are relatively clear. Conversely, developing countries are very broadly defined, encompassing more than 100 countries. Thus, emission reduction obligations are an obligation for a minority of developed countries, which is not conducive to achieving global emission reductions. The refusal of developing countries to join the ranks of those committed to emission reduction has become a pretext for some developed countries to refuse to comply with emission reduction. Therefore, we must recategorize the types of emission-reducing countries based on a comprehensive consideration of emission flows and stocks, per capita and total emissions, and historical and current responsibilities.

Global climate change is a common challenge that is faced by all of humanity, and China is one of the hardest-hit victims of global climate change. As the world's largest coal consumer and sulfur dioxide emitter, and the second-largest energy consumer and carbon dioxide emitter, China has an undisputed duty to cooperate on this issue. China is also one of the countries that is most impacted by climate change. How China deals with global climate change and participates in international climate negotiations is an urgent issue with consequences for the country's future economic and social development. There are two major issues pertaining to China's response to the challenge of global climate change. These are (1) costs and benefits relating to emission reduction and climate change investment; and

(2) strategies and responses relating to participation in international climate change negotiations. These two issues are mutually interactive. However, because of the disproportionate shares of costs and benefits among countries, there are considerable difficulties associated with their responses to the global governance of climate change. Both developed and developing countries should adjust their economic structures to enable them to meet their obligations to reduce emissions. China has the ability and need to actively participate in emission reduction, respond to climate change, and make global environmental contributions.

2.3   Global climate change is our common problem

Because of the uneven costs and benefits for countries, climate negotiations have reached an impasse. Time is of the essence, both for the world and for China, and we have to make a choice in the shortest possible time. About 70%–80% of carbon dioxide emissions are currently generated by developed countries. However, developing countries have also contributed to global carbon dioxide emissions in recent decades. The 2014 *Human Development Report* (UNDP, 2014) noted that the quantities of carbon dioxide emitted in the United States and the United Kingdom in 2010 were 17.6 and 7.9 tons, respectively, reflecting increases of 5.6% and 3.5%, respectively, compared with levels in these countries in 1970. Carbon emissions in China and India were 6.2 tons and 1.7 tons, respectively. The 2007 *Human Development Report* noted that countries with higher levels of human development accounted for the major proportion of total emissions. Moreover, for every 10 tons of carbon dioxide emitted, from the time of the industrial age, about 7 tons have been emitted by developed countries. Historical per capita emissions of the United Kingdom and the United States from 1890 to 2010 are about 1,100 tons, while the per capita levels of China's and India's emissions have been 66 and 23 tons, respectively (see Figure 2.2).

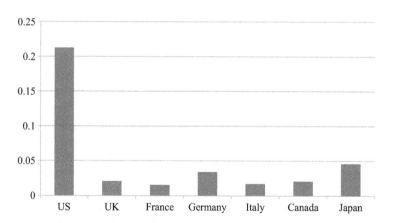

Figure 2.2 The proportion of wealthy countries' carbon dioxide emissions in relation to global emissions (1890–2010)

Source: United Nations Development Programme, 2014 *Human Development Report.*

Table 2.2 Carbon footprints of OECD countries (2012)

	Per capita carbon dioxide emissions (tons)	Carbon dioxide equivalent (million tons)
World	4.51	31,734.00
Australia	16.70	386.27
Canada	15.30	533.74
France	5.10	333.89
Germany	9.22	755.27
Italy	6.15	374.77
Japan	9.59	1,223.30
Netherlands	7.23	173.77
Spain	5.77	266.58
United Kingdom	7.18	457.45
United States	16.15	5,074.14

Source: International Energy Agency *2014 Key World Energy Statistics.*

However, global climate change is a shared problem that affects us all. Whether emissions emanate from New York or Shanghai, the consequences of climate change are the same for everyone. According to the OECD, the earth cannot withstand humans' collective carbon footprint (Table 2.2).[10]

People living in the world's poorest regions are most vulnerable to the impacts of climate change. Thus, the Blair Report also points out that climate change has the greatest adverse impacts on those who bear the least responsibility for it, namely those who are poor and vulnerable. Similarly, poor countries and LDCs that have not significantly increased their GHG emissions, are not currently doing so, and will not do so in the foreseeable future have to bear the costs for climate change. These countries are most vulnerable to climate change, and their ability to adapt is the weakest.

No nation on its own is able to provide GPGs in response to global climate change. Moreover, developing a governance framework that is agreed upon by all countries is not possible based on the institutional mechanisms of the United Nations. The Blair Report expresses concern regarding current global institutional capacities to address climate change. It notes that while the UNFCCC and the Kyoto Protocol provide a sound institutional base, they themselves do not fully meet the requirements for effectively managing the responses and actions needed to tackle climate change. As an international convention, the Kyoto Protocol lacks the necessary penalties and incentives. It only imposes some constraints on a few developed countries, and even these are characterized as "soft constraints," "soft targets," and "soft mechanisms." In reality, only the main powers (major developed and developing countries) are able to take the initiative to provide GPGs, reach a political consensus, and advance and strengthen the new global governance framework.

Currently, GHG emissions are highly concentrated within a few countries. In terms of its emission flow, the United States is the largest emitter, accounting

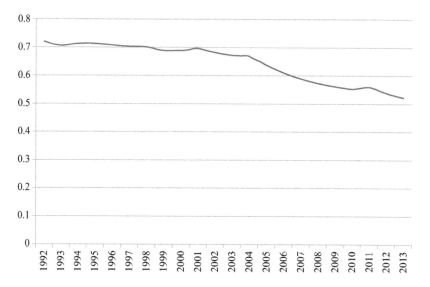

Figure 2.3 The proportion of G8 greenhouse gas emissions to total global emissions (1992–2013)

Source: Authors' calculations based on data compiled from the United Nations and World Bank.

for about one-fifth of the total global emissions. China, the United States, Russia, India, and Japan are the top five emitters, with their emissions collectively accounting for more than half of the total global emissions. However, from 2010 onward, China replaced the United States as the world's largest carbon emitter. The top 10 emitting countries jointly account for 75% of the total global emissions. In the Reference Scenario presented in *World Energy Outlook 2007,* the International Energy Agency (IEA) has projected the growth of GHG emissions entailing a rise in global carbon dioxide emissions by 57% between 2005 and 2030. The United States, China, Russia, and India will account for two-thirds of the increased emissions. In fact, China accounted for the largest share of increased emissions, and India overtook Japan to become the fourth-largest emitter in 2014. From 1980 to 2005, the emissions of the 13 powerful nations (G8+5) accounted for 68%–71% of the world's total emissions. According to the Blair Report, in recent years, this figure has risen to 75%.

According to the Blair Report, commitments obtained just from developed countries are not enough to prevent extreme climate risks; all countries should adjust their economic structures and reduce their carbon dioxide emissions. However, developed countries have made the deepest carbon footprints and should, therefore, take the lead in clarifying their commitments regarding their emission reduction obligations. At the same time, developing countries should also clarify the upper limits of their total emission reductions and their timetables for realizing their emission reduction obligations, based on their circumstances and capabilities.

On July 8, 2008, G8 leaders agreed on a long-term GHG emission reduction target. Accordingly, the G8 nations, together with other contracting states of the UNFCCC, will strive to jointly fulfill at least half of the long-term target for GHG emission reduction by 2050, and will discuss and adopt this target during convention negotiations.[11] However, the "Bali Roadmap" was adopted without legal provisions at the Copenhagen Summit in 2009, thus proving that there is still a long way to go before an effective global climate agreement is reached. In addition, the economic impacts of climate change were initially discussed and proposed at the 2007 meeting of the Group of Twenty (G20) finance ministers and central bank governors of major economic powers. However, "green finance" was first mentioned in a communiqué issued as late as 2016 by participating finance ministers and central bank governors at the G20 meeting held in Shanghai. At the meeting of the Boao Forum for Asia, held in late March 2016, the Chinese foreign minister, Wang Yi, pointed out that in view of the G20 members being mostly large GHG emitters, and also being major participants at the annual United Nations Climate Change Conferences, green finance and climate change funds would be a new focus area to be discussed at the 2016 G20 meeting. We believe that the G20 members need to play a leading role to enable the effective implementation of the Paris Agreement.

The real challenge arises from conversion to a low-carbon economy requiring many interventions at an unprecedented scale, as well as coordination of policies and systems. Therefore, providing financial and technical support is only one aspect of helping developing countries deal with climate change. The top priority is to help developing countries enhance their institutional and policy capacities to tackle climate change. A global governance framework for addressing climate change must provide a combined set of enforceable policies and reference systems for developing countries. However, in the final analysis, developing countries need to improve their ability to adapt to climate change through economic transformation and by strengthening their own capabilities.

In November 2014, the leaders of China and the United States issued a "Sino-US Joint Statement on Climate Change" in Beijing, demonstrating that it is still possible for major world powers to reach a consensus. The period of 2015 to 2020 is a critical one for establishing an international framework convention or global protocol for tackling climate change. China will play a crucial role in breaking the deadlock in climate negotiations and actively promoting good global climate governance. This entails the following: supporting rather than opposing global governance, active rather than passive participation in global governance, commitment to emission reduction obligations rather than attempts to elude them, and taking the initiative rather than taking a free ride to provide GPGs.

2.4 The challenges of addressing climate change brought by the financial crisis

Following the outbreak of the financial crisis, questions ranging from how to deal with it to how countries can stimulate their economies, and to how international economic policies can be coordinated, have dominated discussions at various international forums. In contrast to 2007, when heated discussions on climate change were common, during the second half of 2008 the climate change issue was snubbed. However, after 2012, with the economies of the world's major developed countries continuing to recover, climate change was again put on the global agenda.

Plans formulated by various countries after 2008 to stimulate their economies undoubtedly exacerbated GHG emissions. The financial crisis led to a slowdown of the global economy, and especially of the economies of major energy consumers. This provided sufficient time for politicians to discuss and reach a global emission reduction agreement. However, the economic recovery plans formulated by a number of countries included massive infrastructure projects entailing high energy consumption as well as the relaunching of highly polluting industries. Moreover, economic restructuring was shelved indefinitely, further exacerbating GHG emissions. According to the UNDP's 2014 *Human Development Report,* the world's carbon emissions further increased after 2008. The Stern Report has warned that if no action is taken, GHG emissions will exceed 60 billion tons and 85 billion tons by 2030 and 2050, respectively. Consequently, while a high-carbon economy is reviving, a low-carbon economy is not envisaged in the foreseeable future.

A second impact of the financial crisis is that many countries have lost their ability to respond to climate change investments. To mitigate and adapt to climate change requires large-scale investments, mainly used to enhance countries' abilities to withstand natural disasters, invest in industries that protect the environment, and plant forests to increase carbon sinks. However, as a result of the financial crisis, a number of countries became focused on the competition over financial assistance and technology transfer during international climate negotiations, making it difficult to achieve the desired results from the negotiations. Developed countries on tight budgets have been reluctant to use more funds to help developing countries cope with climate change. In November 2008, the Alliance of Small Island States (AOSIS) called on developing countries to play a leading role in reducing GHG emissions, stating that the current financial crisis should not affect existing international efforts to address climate change. However, this voice has not been widely heard. China's 13th Five Year Plan emphasizes the comprehensive construction of an ecological civilization, and its consequent efforts have brought increasing visibility to the role of developing countries in this regard.

A third consideration relating to the financial crisis is the intensification of its destructive power by climate change. Climate change leads to occurrences of abnormal weather (e.g., snowstorms and drought), causing serious damage in some countries. Following the outbreak of the financial crisis, these countries were subjected to a double crisis. Drought in some countries sparked concerns about food security and led to food price spikes. In the aftermath of the financial crisis, food prices plummeted. Sharp fluctuations in bulk commodity prices further exacerbated the destructive power of the financial crisis. Drought conditions also led to deforestation and increased GHG emissions. In May 2014, Huang Jian-ping and colleagues at the College of Atmospheric Sciences, Lanzhou University, published the latest research findings on climate change in the journal *Nature Climate Change*. These findings indicated that if global GHG emissions continue to increase, arid and semiarid zones across the globe will expand at an accelerated pace, accounting for more than 50% of the entire global land surface by the end of this century. Three-fourths of this expansion of arid and semiarid zones will occur in developing countries.

Leaders of some of the international organizations and politicians of various countries have repeatedly stressed that the process of tackling climate change should, under no circumstances, be delayed because of the financial crisis. This clearly indicates that countries continue to face a serious challenge relating to their efforts to tackle climate change in the context of economic recovery. How-ever, the signing of the Paris Agreement brings new hope and signals a new dawn in relation to the global response to the challenges of climate change after the failure of the Copenhagen Summit to yield any substantive outcomes. Neverthe-less, the cooperation and efforts of the international community are required to effectively implement the commitments and specific targets agreed to by the vari-ous signatories to the agreement.

Notes

1 *G8 Climate Scorecards 2008* was prepared by Ecofys and commissioned by the World Wildlife Fund (WWF) and the Allianz Group. The report compares and ranks the eight countries using nine quantitative indicators, including these countries' emissions trends from 1990 onward and the progress toward implementing the Kyoto targets. At the same time, the report rates the performances of three specific policies: energy efficiency, renewable energy, and carbon markets.
2 China.org.cn, "Fortune 500 Companies Still Increase Carbon Emission, Which Exceed the UN Emission Standards." Accessed December 25, 2014. http://news.china.com.cn/world/2014–12/25/content_34405638.htm.
3 Nicholas Stern, "Stern Review on the Economics of Climate Change," *NBER Working Paper 12741*. www.nber.org/papers/w12741.pdf.
4 Nicholas Stern, *Key Elements of a Global Deal on Climate Change,* 2008. www.lse.ac.uk/GranthamInstitute/publication/key-elements-of-a-global-deal-on-climate-change/.
5 Stern, "Review on the Economics of Climate Change."
6 Stern, *Key Elements of a Global Deal on Climate.*
7 Thomas Schelling, "Addressing Greenhouse Gases," in Nobel Masters toward Energy and Environment-2007 Nobel Laureates Beijing Forum (Beijing: Science Press, 2008), 178–179.
8 Tony Blair, 2008, *Breaking the Climate Deadlock: A Global Deal for Our Low-Carbon Future*. http://blair.3cdn.net/b53ed18eb4812ef5d3_dem6be45a.pdf.
9 Cao Jing, Ling Jing, and Wang Li, "Possible Future International Climate Policy Model – Curbing Global Warming: "Post-Kyoto Era" Where to Go," *Green Leaf,* 5 (2008).
10 "Carbon footprint" indicates the carbon consumption of a person or group. Carbon refers to the carbon atoms composing natural resources such as oil, coal, and wood. The more carbon that is consumed, the more carbon dioxide – the culprit responsible for global warming – will be produced, and the larger the carbon footprint will be. Otherwise the carbon footprint will be small. Wikipedia defines a carbon footprint as

a "measure of the impact human activities have on the environment in terms of the amount of greenhouse gases produced, measured in units of carbon dioxide."

11 G8 leaders believe that the global challenges posed by climate change can only be met if all countries, especially the major economies, contribute to this effort. However, their statement also acknowledges that in accordance with the principle of common but differentiated responsibilities, and the capabilities of individual countries, the actions required to be undertaken by major developed and major developing countries differ. Among the developed countries, eight countries have accepted their leadership responsibility and hold that each member state should implement medium-term targets equivalent to the scale of their own economies so as to achieve absolute emission reductions and curb emission increases as soon as possible. The G8 leaders reconfirmed their commitment to play a leading role in addressing climate change, stating that they would provide technical, financial, and capacity-building assistance to major developing countries to support their emission reduction plans. The G8 leaders also pointed out that to ensure that an effective emission reduction framework is established after 2013, consensus should be reached, internationally, before the end of 2009, whereby all major economies commit, in a binding form, to taking action to mitigate climate change.

3 Global governance of climate change

Reaching a global emission reduction agreement

A number of participating countries in the global governance of climate change face the dual problem of non-competitive costs with non-competitive benefits in emission reduction. They also face the challenge of ensuring equitable use and allocation of public resources while maximizing national interests and reducing the cost burden. The costs of reducing global emissions may not be high; however, because of variations between countries and their policies, emission reduction measures tend to have lower economic impacts if they are implemented sooner. From the perspective of fairness and efficiency, we believe that countries at different stages of their development, and with different levels of emission reduction, should have commensurate responsibilities and obligations. We have followed two principles in classifying countries according to their global emission reductions: the Human Development Index and major emitters. Thus, we have creatively replaced the traditional classification method, which is based on the two categories of developed and developing countries, with one entailing four groups based on the Human Development Index. Additionally, we have provided binding targets for global emission reductions. As the world's largest developing and developed countries, respectively, and the two largest CO_2 emitters, China and the United States should adopt a more positive and pragmatic attitude and make greater contributions in terms of international cooperation in addressing the challenges of climate change.

3.1 Losses caused by global emissions

According to the latest scientific research presented in the 2014 report of IPCC's Working Group III, to ensure a 66% probability of maintaining a GHG reduction level with a global temperature rise of no more than 2 °C necessitates reductions in global consumption of 1%–4%, 2%–6%, and 3%–11% in 2030, 2050, and 2100, respectively. This estimation does not account for some of the other associated benefits such as improved air quality brought about by emission reduction, and is likely to have underestimated progress related to low-carbon technological innovations introduced after 2030. Thus, despite the fact that considerable controversy exists over estimations of future emission reduction costs, a significant number of economists believe that the results of these estimates, based on different models and entailing a variety of assumptions, are likely to have greatly overestimated economic losses caused by emission reductions. The Blair Report[1] contends that the impacts of emission reduction on economic growth may be even lower than those reported by previous studies. It supports this argument with the following observations.

- On average, the United States spent about 6% of its annual GDP on defense during the Cold War (1950–1990).
- About 3% of the global annual GDP is spent on insurance.
- The estimated global cost of the subprime mortgage crisis in the United States, for the financial sector, potentially accounted for 2% of GDP.
- The cost of the global petroleum price increase from US$40/bbl to US$130/bbl between June 2004 and June 2008 amounted to about 5% of the GDP.

In a scenario of normal economic growth, the per capita GDP in 2050 is projected to increase from the current US$5,900 to US$15,900. However, in a low-carbon economy, the projected per capita GDP in 2050 will range between US$15,000 and US$15,600. Thus, for a scenario entailing a low-carbon economy, income levels will be 2.5 times higher than current rates (see Figure 3.1).

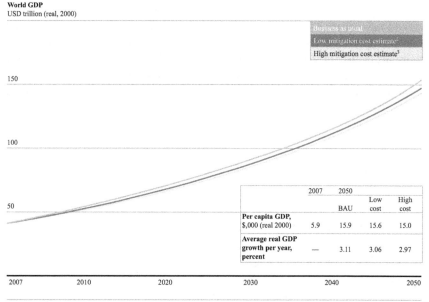

World GDP
USD trillion (real, 2000)

Business as usual
Low mitigation cost estimate[2]
High mitigation cost estimate[3]

	2007	2050		
		BAU	Low cost	High cost
Per capita GDP, $,000 (real 2000)	5.9	15.9	15.6	15.0
Average real GDP growth per year, percent	—	3.11	3.06	2.97

1 Global Insight GDP forecast to 2037, extrapolated to 2050
2 1% of GDP in 2030, 1.8% in 2050
3 3% of GDP in 2030, 5.5% in 2050

Figure 3.1 The proportion of the world GDP emission reduction cost relative to the growth rate (2007–2050)

Source: Tony Blair, "Breaking the Climate Deadlock: A Global Deal for Our Low-Carbon Future," 2008, 28.

According to calculations made by the International Monetary Fund (IMF), concerted efforts to mitigate climate change can produce rapid and widespread macroeconomic impacts. However, by gradually increasing the carbon price from a low starting point at an early stage, the cost of adjustment can be shared over a longer period so as to reduce it to a minimum. If, as a global community, we had begun to implement a policy for emission reduction in 2013, and continued to strive to stabilize the concentration of CO_2 equivalent at 550 ppm (a concentration of one part per million) by 2100, the net present value of world consumption in 2040 would evidence a decrease of only 0.6%. Notwithstanding this loss, the world's gross national product in 2040 would still be 2.3 times higher than it was in 2007. IMF estimates indicate that the global cost of emission reduction may be low during the period from 2013 to 2040. However, the cost will vary across different country and policy contexts. The cost amount and its distribution among countries are very sensitive to the ease with which countries can reduce their own emissions and their formulation of specific policies relating to mitigation impacts.[2]

The findings of a study conducted by William Cline, a senior fellow at the Peterson Institute for International Economics and the Center for Global Development, has revealed that almost all countries are affected by climate change.[3] Overall, developing countries will suffer greater agricultural losses than industrialized countries unless they actively respond to climate change. The loss and gain in China's agriculture are both 7%. Thus, China's interest in emission reduction is higher. India will potentially face considerable losses in the event of any policy change relating to global emissions. Cline suggests that because most of the developing countries are at risk, they are strongly profit-driven and motivated to join the global emission reduction plan. Consequently, he believes that China is more interested in promoting emission reduction. According to our preliminary calculations, given that China's potential growth rate is expected to slow down in the coming decades, the annual emission reduction cost will have little effect on national economic development. According to the data provided in the Blair Report, China's GDP is expected to grow to 6.9 times its current value by 2040, *even in a scenario entailing the development of a low-carbon economy. In a scenario entailing no emission reduction, China's GDP is projected to grow by 7.2 times*. This calculation does not take into account the huge gains for China from reducing emissions and converting to a low-carbon economy.

Development of a low-carbon economy helps to promote employment. The process of establishing a new low-carbon economy offers opportunities for the pooling of major investments, creating jobs and generating numerous business opportunities. For example, the value of global renewable energy technology subsidies reached US$120 billion in 2013. By 2040, it is expected that the capacity of electricity generated from renewable energy will comprise nearly half that of the world's newly generated electricity, entailing a threefold increase in the consumption of biomass fuels up to 4.6 million barrels/day. On a global scale, wind power generation capacity accounts for the largest proportion of renewable energy growth (34%), followed by hydropower (30%) and photovoltaic power (18%). Within the European Union, wind power generation capacity accounts for 20% of the total power generation capacity, and in Japan, the proportion of photovoltaic power generation capacity reaches 37% of its total power generation capacity during the peak summer period.[4] According to the *Renewable Energy and Jobs – Annual Review 2015* issued by the International Renewable Energy Agency (IRENA), there were about 7.7 million people employed globally within the renewable energy industry in 2015, compared with 6.5 million employed in this industry in 2014, indicating an increase of 18%. Of these employees, 3.4 million (40%) were based in China. Moreover, the IRENA predicts that by 2030, the share of renewable energy will double in the global energy sector as a whole, with the creation of over 16 million jobs worldwide. The Blair Report proposes an institution and mechanism for implementing actions, and expresses full confidence in the role of the global carbon trading market. An effective global carbon trading market would succeed in lowering the cost of reducing emissions by 50%. Because the cost of transitioning to a low-carbon economy would be shared annually, the cost of emission reduction would be lower than anticipated. According to data presented in the Blair Report, the marginal cost of developing a low-carbon economy for the United States would be about $1.1 trillion, which is just 1.5% of the total investments of US$77 trillion envisaged during the same period. Most of the costs associated with the establishment of a low-carbon economy can be repaid gradually over time. Moreover, predicted costs have historically often exceeded actual costs, as it is difficult to predict responses relating to incentive measures such as the price of carbon or the pace of technical innovation. For example, in 1988, American economists predicted that the cost of halving chlorofluorocarbons (CFCs) was US$21 billion. However, by 1990, following two years of implementation of the Montreal Convention, the cost of totally eliminating CFCs had come down to US$2.7 billion, which was 87% lower than the predicted cost for halving CFCs.[5]

However, there are significant variations in emission reduction costs among countries. *A Cost Curve for Greenhouse Gas Reduction,* published by the McKinsey Global Institute, suggested that a 450-ppm scenario for the global economy will depend on countries' capacities to implement all available measures to reduce GHG emissions at a cost of 40 euros per ton. If this happens, the cost curve shows that by 2030, the global annual cost may rise to about 500 billion euros, representing 0.6% of the projected GDP. However, if higher price measures are needed to achieve the emission reduction targets, the cost may further rise to 1.1 trillion euros, accounting for 1.4% of the global GDP. However, for developing countries, the actual cost of reducing emissions is much greater than the direct costs associated with the technology.[6] In their report *China's Green Revolution,* issued in early 2009, McKinsey & Company showed that fully exploiting technologies with the greatest potential could not only significantly improve China's energy security situation, but could also control GHG emissions at a level of about eight billion tons in 2030, which is only about 10% higher than the level in 2005. However, to achieve that potential requires substantial new investments. This report estimates that over the next 20 years, China will, on average, increase new investments by 150–200 billion euros annually (equivalent to about RMB 1.3–1.7 trillion). One-third of this investment can generate economic returns, one-third can generate low to moderate economic costs, and one-third can generate huge economic costs. In other words, achieving the envisaged emission reduction scenario requires efforts on the part of China that amount to nothing less than a "green revolution."[7]

3.2 Establishment of an international mechanism for addressing climate change and achieving collective action

The world's major economies are bound to promote the provision by all countries of GPGs in response to climate change. This is also important for promoting good governance regarding global climate policies and breaking the current deadlock in international climate negotiations. Kaul, Grunberg, and Stern (1999) have argued that from a national perspective, the beneficiaries of GPGs are not just a particular nation; they extend to several groups or even to the entire population of a country. Similarly, from a generational perspective, not only the present generation comprises the beneficiaries, but also a few succeeding generations, or at least several contemporaneous generations, without adversely impacting on future development options of several future generations.[8] The classification of GPGs presented by Kaul et al. (see Table 3.1) further reveals that both benefits and costs for addressing climate change are non-exclusive. In other words, all countries will enjoy the benefits of improvements relating to climate change, or they will assume the costs of any climate change–related deterioration. However, because the costs are non-competitive, with none of the benefits being non-competitive, while all of the countries would be willing to share the benefits, they would conversely be reluctant to bear these costs. This is the underlying cause of the prisoner's dilemma in international climate negotiations.

In the absence of a world government, GPGs are often provided by hegemonic countries, by relevant international organizations, or through international agreements. For example, the IMF provides financial stability, the World Trade Organization (WTO) provides free trade, and the World Bank provides freedom from poverty, equality and justice, and other GPGs through long-term loans offered to developing countries. The World Bank provides loans and technical assistance to member governments or private enterprises guaranteed by governments to enable them to develop construction projects within a specified time frame, and with low profit margins, which are required for a country's economic and social development. GPG providers may also be international laws or conventions rather than organizations. Participation of the contracting parties, and their recognition that the international law or convention has entered into force, is required; otherwise the latter cannot assume the function of providing GPGs. Some providers of GPGs may also comprise multiple international organizations or international conventions. Examples of such organizations that address climate change include

Table 3.1 Classification of global public goods

Class and type of global good	Benefits		Nature of the supply or use problem	Corresponding global harmful goods	Costs	
	Non-excludable	Non-rival			Non-excludable	Non-rival
1. Natural global commons providers: None.						
Ozone layer	Yes	No	Overuse	Depletion and increased radiation	Yes	Yes
Atmosphere (climate)	Yes	No	Overuse	Risk of global warming	Yes	Yes
2. Human-made global commons providers: examples include Human Rights Organization, the World Intellectual Property Organization, Intellectual Telecommunication Union, and the International Organization for Standardization.						
Universal norms and principles (such as universal human rights)	Partly	Yes	Underuse (repression)	Human abuse and injustice	Partly	Yes
Knowledge	Partly	Yes	Underuse (lack of access)	Inequality	Partly	Yes
Internet (infrastructure)	Partly	Yes	Underuse (entry barriers)	Exclusion and disparities (between information-rich and information – poor countries)	Partly	Yes
3. Global conditions providers: examples include the UN Security Council, World Health Organization, International Monetary Fund, World Trade Organization, and the World Bank.						
Peace	Yes	Yes	Undersupply	War and conflict	Partly	Yes
Health	Yes	Yes	Undersupply	Disease	Yes	Yes
Financial stability	Partly	Yes	Undersupply	Financial crisis	Yes	Yes
Free trade	Partly	Yes	Undersupply	Fragmented markets	Yes	Yes
Freedom from poverty	No	No	Undersupply	Civil strife, crime, and violence	Yes	Yes
Environmental sustainability	Yes	Yes	Undersupply	Unbalanced ecosystems	Yes	Yes
Equity and justice	Partly	Yes	Undersupply	Social tensions and conflict	Yes	Yes

Source: Derived from Inge Kaul, Isabelle Grunberg, and Marc Stern, *Defining Global Public Goods, in Global Public Goods: International Cooperation in the 21st Century.* The authors have included different providers of global public goods in accordance with the functions of the relevant international organizations.

the Intergovernmental Panel on Climate Change (IPCC), the World Meteorological Organization (WMO), the UN Security Council, the EU, and the Association of Southeast Asian Nations (ASEAN), among many others. However, not all of these international organizations are intergovernmental. There are also a number of non-governmental international organizations.

On the issue of responding to global climate change, hegemonic countries are unable to individually provide GPGs. These must be advocated and promoted jointly by a number of powers. Currently, the main international organizations and institutions focusing on climate change are the WMO and the IPCC. However, these two institutions lack the necessary mechanisms to encourage countries to engage in collective actions. The WMO's predecessor was a non-governmental organization, the International Meteorological Organization, founded in 1873. Based on an agreement reached with the UN, the WMO was officially reconstituted as a specialized UN agency in 1951. This organization has the following aims: to promote the establishment of a worldwide meteorological observation network; to facilitate rapid international exchange of meteorological data; to implement standardized operational meteorological observations; to propose unified published specifications for observations and statistics; to promote the application of meteorology in aviation, navigation, water resources, and agriculture; to promote the implementation of hydrological operations; to strengthen cooperation between the meteorological and hydrological services sectors; and to encourage scientific research and training of personnel in meteorology and related fields. The IPCC is an intergovernmental body jointly established by the WMO and the United Nations Environment Programme (UNEP) in 1988. Its main tasks are to assess the status of scientific knowledge on climate change, as well as the potential social and economic impacts of climate change, and possible adaptation and mitigation countermeasures. Its mandate is solely to provide decision makers with relevant information on climate change. The agency itself does not conduct any scientific studies. Rather, it reviews thousands of annually published papers on climate change and releases an assessment report every five years that summarizes "existing knowledge" on climate change.[9] To provide GPGs that address climate change, it is evident that specific international mechanisms or revamped existing international organizations are required, along with the establishment of a mechanism to provide incentives and penalties so as to promote collective action on a global scale.

Stern holds the same view. On October 23, 2008, when he delivered a speech on China's economic and climate change at Tsinghua University's School of Public Policy and Management, he stated:

A British, Maynard Keynes, and an American, White, jointly designed the "Bretton Woods system" in the 1940s, and they designed what they called the "Three System" organizations: World Bank, IMF and WTO. If they sit down again today to discuss it, I think they will redesign the systems on the environment-related topic as I mentioned just now. I think they will combine the World Bank and the IMF, and may set up a world trade organization and an environmental organization. Major change has occurred since the 1940s in the world. One part is the environmental issue, and the other part is the balance of international affairs and economic power, Therefore, China will become and should join the center for negotiating these three issues on the world stage: economy and trade, climate, and environment.

Stern fully expects China to play a decisive role in building a future international institution. He believes that China, as a WTO member, should align its actions to be consistent with the original objectives of the organization. China will play an important role and effect change in specific matters during the course of trade negotiations and consultations during the development of a new global financial system. Given the size of its economy, and its growth rate, production methods, and the demand for energy and commodities, China will profoundly alter the global economy. It will also play an important role in transforming the components of global economic and financial systems. China's impacts on the World Bank's and IMF's reforms in relation to global economic systems should be far-reaching. Moreover, the measures to be taken by China will also significantly influence the global position on climate change and environmental issues.[10]

Thomas Schelling proposed a new organizing principle, other than the Kyoto Protocol, consisting of two parts. First, participation should be limited to those countries that are serious, united, and cooperative and should not extend to all 192 countries. Second, developing countries must heed their own development when reducing the effects of climate change and cannot address climate change at the expense of their future. Schelling's proposed program entails developing a common international mechanism in times of peace, with all participating countries declaring their commitment and following these assiduously. However, such "self-respecting" countries do not have to commit to carbon emission caps. They just need to promise to reduce carbon emissions and take actions such as raising taxes, increasing subsidies, and enhancing research efforts.[11] Schelling's view represents the view of many economists, and we too basically subscribe to it. It resonates with our view that as a large developing country, China needs to commit to reducing emissions and taking feasible actions.

3.3   Cooperatively addressing climate change as the best choice for all countries

To address global climate change and achieve good governance through countries' national climate policies requires all countries to collectively respond to this issue through cooperation, initially among the large powers. The time frame for the cooperation of large powers in promoting global governance for addressing climate change is a very tight one, and the cost of governments' non-action is high. A UNDP report notes that in the seven years since the Doha Round started, to continue the analogy, stocks of greenhouse gases have increased by around 12 ppm – and those stocks will still be there when the trade rounds of the 22nd century get underway.

A country's participation in international negotiations relating to climate change is linked to its concerns on how to maximize its interests and reduce cost burdens in the equitable use and allocation of public resources. The UNDP has recognized this issue, noting that the most difficult policy challenges will relate to distribution. While there is potential catastrophic risk for everyone, the short- and medium-term distribution of the costs and benefits will be far from uniform. The distributional challenge is made particularly difficult because those who have largely caused the problem – the rich countries – are not going to be those who suffer the most in the short term. It is the poorest who did not and still are not contributing significantly to greenhouse gas emissions that are the most vulnerable.

However, when it comes to measuring the responsibilities to be borne by different countries with different standards, conclusions will differ.

If we examine the emission trend, we will find an increasing similarity in emissions from developed and developing countries, with the proportion of developing countries responsible for global emissions steadily rising. In 2012, the amount of CO_2 emissions worldwide was 31,734.3 Mt, which showed an increase of 51% compared with the amount of these emissions in 1990 (20,973.9 Mt). In 2012, the total amount of CO_2 emissions from fuel combustion contributed by the top 10 emitters was equivalent to nearly two-thirds of the total amount of global CO_2 emissions. As the world's largest emitters, China and the United States were responsible for 8,250.8 Mt and 5,074.1 Mt of CO_2, respectively, in 2012, accounting for 26% and 16% of global emissions, respectively. These quantities far exceed those of other countries. India is the third-largest emitter, and even though it accounts for only 6% of global emissions, its emissions are increasing at a rapid pace.[12] The proportion of CO_2 emissions from OECD countries in relation to total global emissions has continued to decline from 66.1% in 1973 to 38.3% in 2012. It is expected that by 2035, the proportion of CO_2 emissions from these countries will be only 27.3% of total global emissions.[13]

If we also account for deforestation, global rankings for CO_2 emissions will change. If the World Rainforest is considered as a "country," this "country" would rank highest for global CO_2 emissions. If we only consider emissions attributed to deforestation, Indonesia would be the third-largest source of annual CO_2 emissions (2.3 billion tons of CO_2 emissions annually), with Brazil ranking fifth (1.1 billion tons of CO_2 emissions annually).

Countries usually sign treaties that are advantageous in terms of their own interests (rather than those of other countries).[14] Thus, countries would be expected to choose standards that are in their favor and to seek to maximize their national interests during international climate negotiations. However, politicization of scientific issues leads to endless debate during these negotiations. As a result, no pragmatic, feasible, and effective measures are introduced, and humanity, especially the majority of the poor, will consequently have to pay a higher price.

Challenges relating to energy, the environment, and climate change that currently face countries throughout the world cannot be solved under the existing framework of global governance. It is only through cooperation that the current crisis can be avoided. Zhang Huanbo (2007) used a multi-country climate protection simulation model to examine three scenarios: no implementation of emission reduction, implementation of cooperative emission reduction, and implementation of non-cooperative emission reduction. The findings of the study revealed that in the absence of an international agreement on emission reduction, no country would take the initiative to reduce emissions, and the effectiveness of emission reduction under these circumstances would be worse than it would be under conditions of cooperation. Therefore, a strategy entailing global cooperation to address climate change is the most effective.[15] We believe that the key to achieving cooperation is to identify a widely shared interest that is acceptable to all the parties.

3.4   Two classification principles for countries participating in global emission reduction

Considering the emissions of countries as an issue requiring global action, we can classify over 200 countries worldwide into four groups based on principles of fairness and efficiency rather than relying on the traditional division into two groups. Consequently, we can determine appropriate emission reduction contributions according to the quantities of emissions emanating from the major emitters in proportion to the total amount of global emissions. Accordingly, we posit two principles, described below.

The first principle entails the replacement of a binary division of developed and developing countries with a division of countries into four groups based on their HDI.

From 1990 onward, HDI has been widely applied as an important indicator for measuring human progress. The index is calculated based on life expectancy at birth, years of education (including the average number of years of schooling and the expected number of years of schooling), and per capita gross national income. Based on these criteria, countries can be compared. A total of 187 countries were classified into four groups according to their HDI values presented in a report released by the UNDP in 2014: an extremely high level of human development, a high level of human development, a medium level of human development, and a low level of human development. The basic classification procedures were as follows:

- Measure the basic impacts of human development.
- Include only limited variables to facilitate calculation and management.
- Use a composite index rather than too many independent indicators.
- Include both economic and social options.
- Maintain flexibility of the index range and theory.
- Provide guarantees for highly credible data sources.

HDI classification is based on cut-off points on the quartile of the HDI component index as follows: extremely high HDI (HDI ≥ 0.800), high HDI ($0.700 \leq$

HDI < 0.799); medium HDI (0.550 ≤ HDI < 0.699), and low HDI (HDI < 0.550). These groupings can be envisaged as "One Earth, Four Worlds." According to the UNDP's 2014 *Human Development Report,* only the extremely high HDI countries belonging to the first group can be developed countries (these include countries such as Qatar and the United Arab Emirates). The remaining three groups comprise developing countries.

Currently, the extremely high HDI group comprises 49 countries, accounting for nearly 1.2 billion people (about 17% of the world's total population). These countries are targeted as those that should unconditionally reduce their emissions and should follow the relevant emission reduction principles of the UN.[16] An increasing number of countries will enter the extremely high HDI group. According to data for the period from 1980 to 2013 on HDI trends contained in the UNDP's *Human Development Report 2014,* the extremely high HDI group comprised 40 countries and regions in 2005, implying that unconditional emission reductions will apply to an increasing number of countries or regions.

The high HDI countries (including China) collectively account for nearly 2.5 billion people, or about 35% of the total global population. They should constitute a second tier of conditional emission-reducing countries. Here, "conditional" is defined by the gap between a country's HDI index and the value of 0.8 (the lowest value of an extremely high HDI country); the narrower the gap for a particular country, the higher its share of emission reduction. When a country's HDI within its current group (greater than or equal to 0.8) reaches the level of the extremely high HDI group, its status will change from being a conditional to an unconditional emission reduction country. For example, China's HDI was 0.719 in 2014. When, in the future, it reaches or exceeds 0.8, it will be recategorized as an unconditional emission reduction country.

The UN should set up a special agency to monitor countries' emission reduction measures and the impacts of unconditional and conditional emission reduction countries. These findings should be regularly published.

Currently, there are no provisions on the mandatory emission reduction obligations of the remaining two groups, namely medium and low HDI countries. However, these countries should be urged to implement voluntary emission reduction measures according to their domestic conditions (see Table 3.2).

The second principle is the major emitters' emission reduction. Currently, the top 20 emitters account for nearly 80% of the world's total emissions (Table 3.3). They are both major emitters and major emission reducers. If a country's emissions constitute a high proportion of total global emissions, this indicates that its contribution to global emission reduction should be commensurately high. Countries' emission reduction credits are shared in accordance with their negative externalities in relation to global emissions. Those with higher emissions will thus have higher emission reduction obligations or credits. China and the United States should take the lead in emission reduction as these two countries together account for 42.33% of total global emissions, followed by countries such as India, Russia, and Japan. The respective emission shares of these latter countries are close to or above 4%, each, totaling 15.42%, while the emissions of the remaining

Table 3.2 "One Earth, Four Worlds"

Category (HDI range)	Groups (Based on HDI)	Emission reduction conditions	Countries		Population	
			Qty. (pcs)	Ratio (%)	Qty. (millions)	Ratio (%)
0.800–1.000	Extremely high HDI	Unconditional emission reduction	49	26.20	1,189.7	16.61
0.700–0.799	High HDI	Conditional emission reduction	53	28.34	2,485.5	34.70
0.550–0.699	Medium HDI	Recommended emission reduction	42	22.46	2,262.1	31.58
< 0.550	Low HDI	Recommended emission reduction	43	22.99	1,145.6	16.00

Table 3.3 Top 20 CO_2 emitting countries in the world (2012)

Rank	Country	Total emission (Mt CO_2)	Proportion of global emissions (%)	Human Development Index (HDI)
1	China	8,205.86	26.15	0.715
2	United States	5,074.14	16.17	0.912
3	India	1,954.02	6.23	0.583
4	Russian Federation	1,659.03	5.29	0.777
5	Japan	1,223.30	3.90	0.888
6	Germany	755.27	2.41	0.911
7	Korea	592.92	1.89	0.888
8	Canada	533.74	1.70	0.901
9	Iran	532.15	1.70	0.749
10	Saudi Arabia	458.80	1.46	0.833
11	United Kingdom	457.45	1.46	0.890
12	Brazil	440.24	1.40	0.742
13	Mexico	435.79	1.39	0.755
14	Indonesia	435.48	1.39	0.681
15	Australia	386.27	1.23	0.931
16	South Africa	376.12	1.20	0.654
17	Italy	374.77	1.19	0.872
18	France	333.89	1.06	0.884
19	Turkey	302.38	0.96	0.756
20	Poland	293.77	0.94	0.833
	Total	24,825.39	79.13	
	World	31,374	100.00	0.700

Source: IEA, *2014 Key World Energy Statistics;* UNDP, *Human Development Report 2014.*

15 countries account for 21.38% of total global emissions. This principle is similar to that underlying the dynamic map of countries participating in emission reduction. We should also consider the HDI levels of different countries. Among the 20 major emitters, 11 countries belong to the extremely high HDI group and

are categorized as unconditional reducers and 6 countries belong to the high HDI group and are categorized as conditional reducers. The remaining three countries, namely India, Indonesia, and South Africa, belong to the medium HDI group. However, as the third-largest emitter in the world, India should actively implement emission reductions. When India enters the high HDI group, it will correspondingly be recategorized as an active conditional reducer.

It should be noted that a classification principle entailing a division into four groups based on HDI could also be usefully applied in relation to future global payment transfers. Thus, extremely high HDI countries would become major aid-giving and technology-transferring countries; low HDI countries would directly obtain Official Development Assistance (ODA) and free or low-cost transfer of technology; medium HDI countries would obtain low-interest loans from international financial organizations, as well as low-cost transfer of technology; and high HDI countries would obtain assistance in transferring technology. Annual HDI data published by the UNDP on all countries would constitute the main information base for developing global emission reduction programs and providing economic assistance, as these data are both simple and transparent.

These two principles can be used to develop binding indexes for global emission reduction. The quantity of a country's emission reduction is determined by two sets of factors. The first relates to the stage of the country's development, which includes its emission stocks, per capita emission levels, and historic responsibility. As HDI is a very good measurement index that is preferable to the per capita GDP grouping index, it should serve as the primary measurement index. The second relates to a country's total contribution to global emissions, flows, and actual responsibility. The top 20 emitters directly influence the realization of global emission reduction targets and actions. Consequently, their future emission reduction credits should be directly linked to total emissions. We considered applying the above principles to develop a roadmap for countries (mainly those with high HDI and the 20 major emitters) to reach their emission reduction targets by 2050. This entailed clearly defining the different categories of countries and determining their responsibilities and obligations for reaching a global emission reduction agreement.

3.5   President Obama's new energy deal and prospects for a global emission reduction agreement

Obama does not regard the struggle to tackle climate change as a burden placed on the United States. Instead, he believes that the response to climate change and efforts to revitalize the American economy are not contradictory. Rather, he holds that the development of clean energy is part of these efforts to revitalize the American economy, which can create millions of jobs and provide a new focus for economic growth. It is evident that the purpose of some of Obama's most important energy policies to date is to stimulate the economy and promote employment. For example, during the initial stage of his tenure, Obama proposed the goal of investing US$150 billion during the next 10 years with the aim of stimulating the private development of new clean energy and ensuring that the proportions of electricity generated in the United States from renewable energy sources reached 10% and 25% in 2012 and 2050, respectively, while the number of plug-in hybrid vehicles reached one million by 2015. Although this policy would not have a significant immediate effect on energy costs, the number of jobs created within the renewable energy industry would be considerable. The creation of new green jobs includes the following benefits, in addition to increasing the proportion of power generated from renewable energy sources: the use of the cheapest, cleanest, and most efficient energy resources; improving the performances of the homes of one million families a year in providing resistance against the winter cold; development and use of clean coal technology; and construction of the Alaska natural gas pipelines. Evidently, then, Obama's plan to stimulate the economy in response to the financial crisis and programs to address climate change are not contradictory, and can even prove to be all-in-one solutions.

The biggest obstacle to Obama's energy policy comes from industrial interest groups, particularly car manufacturers. However, car manufacturers in the United States faced bankruptcy during the financial crisis, compelling them to obtain government loan assistance. They have had to compromise with the government, and shift their focus to the future development of fuel-efficient cars. This developmental direction exactly matches that of Obama's energy policy. New technologies, as well as protection through relevant legislation for its large-scale promotion, will result in the gradual reduction of long-term costs, with technology becoming key to the American auto industry's competitiveness.

In fact, Obama's "New New Deal" had already been implemented on a "trial and error" basis by state-level governments in the United States. Moreover, it resembles much of the institutional innovation and opening up at the level of local governments under China's reform. Obama merely endorsed these system-related innovations and authorized them at the level of the federal government. Although previous federal governments of the United States had long held a negative attitude toward emission reduction, state and municipal governments have done much on the environmental protection front and are far ahead of the federal government in this regard. For example, in September 2008, 10 northeastern states launched the first mandatory "cap-and-trade" program relating to CO_2 emissions. This "Regional Greenhouse Gas Initiative" program limited the total permitted CO_2 emissions from these states' power-generating industries in 2014, with the further requirement to reduce their total emissions by 10% in 2018. More than 800 mayors from all 50 American states have sworn that their cities would meet or exceed the Kyoto targets. Thus, the "New Deal" on energy and environment launched by Obama was not actually new in the United States, and a solid policy foundation and considerable experience already existed in the country.

The evidence on climate change has repeatedly proven that the longer we delay acting on this issue, the graver the situation faced by all of humanity will become. According to a Gallup poll conducted in 2014, 41% of Americans believe that the impacts of climate change have been exaggerated. This indicated a decrease by 7% compared with the percentage of Americans holding this belief in 2010. At the same time, the number of people who believe that environmental issues are real, while underestimated, has shown a gradual increase. In acknowledgement of this situation, and bypassing Congress, in June 2014, the Obama administration launched a massive and mandatory reduction decree in the hopes of reducing levels of CO_2 emissions from power plants by 2030 to 70% of their levels in 2005. Not surprisingly, this immediately fueled fierce criticism from Republicans who believed that Obama's proposal would not only stifle employment within the coal industry, but would also increase the cost of electricity. In November 2014, China and the United States jointly issued the "Sino-US Joint Statement on Climate Change," whereby the United States committed to reducing its emissions by a maximum of 28% by 2025, based on its carbon emission levels in 2005. Various surveys have revealed the American electorate's position on the issue of climate change. However, Republicans remain convinced that efforts to further reduce carbon emissions will result in the unemployment of Americans, thus threatening the long-term growth of the country's economy.

The United States is the top-ranking power in the world, but its environmental policies change dramatically with a change of government. This can be attributed not only to the personal understanding and thinking of the new leader, but also to deeply embedded factors relating to the country's economy, as well as the areas of diplomacy and security. The United States is a country oriented toward pragmatism. However, positive changes relating to the climate and environment, introduced by the new government, do not imply a desire on its part to act as a savior as a result of a sudden awakening of conscience. Rather, the aim is to safeguard the next phase of the economy, diplomacy, and the security interests of the United States, while maintaining worldwide leadership in relation to the new domestic and international environments. An understanding of the underlying reasons for these policies and actions has very important implications for China during the next global exchange on climate issues and the formulation of domestic environmental strategies.

Obama's energy and environment policy has very real implications for diplomacy and security. Currently, the United States has not fully withdrawn from Iraq and Afghanistan, and the Iranian nuclear issue, the Palestinian-Israeli conflict, and other regional instabilities have become grave. It is becoming increasingly unrealistic for the United States to maintain its long-standing position of trying to achieve complete control in the oil-producing regions of the Middle East, posing a serious threat to America's oil and energy security. Thus, energy security and environmental protection are at the core of Obama's "New Energy Plan."

The United States has long been the largest energy consumer and importer in the world, and the country's economy is strongly dependent on oil. Its combination of energy sources in 2013 comprised fossil fuels (81%) and nuclear power (8%), with renewable energy (hydropower, geothermal, solar, and wind) accounting for 8%, with biomass at just 1%. Among the fossil fuels, oil was the major component (36%), followed by natural gas (27%) and coal (18%).[17]

The financial crisis in 2008 resulted in a significant drop in international oil prices from the historically high price of US$147/bbl to around US$40/bbl. With the slow recovery of the global economy, the price of oil gradually rose to nearly US$100/bbl in 2012, and remained at about US$90/bbl over the next two years. In mid-2014, because of the rapid growth in crude oil production in the United States, as well as continuous increases in inventories and in the dollar exchange rate, international oil prices fell from nearly US$100/bbl to US$50/bbl. However, it is clear that the price of oil cannot remain at this low level for long. *Energy Outlook 2014,* issued by the US Energy Information Administration (EIA), anticipates that international oil prices will reach US$119/bbl (at the dollar value in 2012) by 2030. Obama believes that the United States relies excessively on imported oil, which is a threat to its economy and its national security. Consequently, the country's energy policy requires adjustment. In his inaugural speech, Obama pointed out: "The way we use energy strengthens our adversaries and threatens our planet." In recent years, a number of oil-producing countries have changed or are ready to change the oil trade settlement currency. This would seriously threaten the hegemony of the dollar. The performances of Venezuela, Russia, and other countries in recent years have also heightened awareness among American decision makers of the serious threat posed by inflated oil prices and overreliance on oil imports to the national security and global strategies of the United States.

Obama's "New Energy Plan" entails several major policies. These include improving energy efficiency and reducing dependence on imported oil, particularly fossil fuels; emphasizing the development of alternative energy sources, increasing investments, and promoting research and the application of new energy-related technologies; and stipulating the power capacity from renewable energy, improving fuel economic standards, and reducing carbon emissions. These policies will reduce the dependence of the United States on oil in the long run, increase the safety factor of its energy use, and enhance its ability to control the oil-producing countries. From a more in-depth perspective, the green energy policy of the United States entails aggressive, far-reaching, and long-term strategic considerations. A Chinese scholar, Wu Jiandong, believes that it is even more significant than Bush's direct attack on Iran and Iraq, and would be a drastic blow to the consortium of oil producers, which includes Russia as well as countries in western Asia and South America. It will compel them to abandon their self-interest in maintaining high oil prices and thereby revert to the hierarchical structure of the global economy, which is at a much lower level than that of the United States. This will consequently lead to the restructuring of the global power structure. Change in the energy structure will completely change the pattern of world energy, thereby changing the global economic and political situation.

Obama's team clearly recognizes the economic, political, and strategic significance of changes in energy use patterns, which is reflected in the energy policy. It can be said that, represented by Obama, contemporary political leaders are steering an unprecedented industrial revolution that will change the entire global situation. Moreover, the starting point of this revolution is transforming energy use patterns.

Obama's energy and environment policies thus have far-reaching international impacts. Evidently, Obama's aggressive environmental policy will facilitate the progress of the international emission reduction agreement. The participation of the United States provides huge impetus for the engagement of both developed and developing countries.

European countries are keen to reduce their emissions and aspire to seize the high ground of morality and international rule-making power. However, it is clear that Europe cannot replace the United States as a global leader and major international rule-maker. The long-standing evasive attitude of the United States on this issue has created frustration within these countries, which also face domestic pressures. Most people are concerned that their respective country's contribution to emission reduction will be offset by increasing emissions from other countries. That the United States has now joined the ranks of the world's emission-reducing countries will undoubtedly greatly encourage the European countries, and will serve to further strengthen and promote the decisions and actions of developed countries in continuing to reduce their emissions.

We should not completely ignore the fact that the long-term targets of the United States on GHG emission reduction and those of the international community are basically the same: namely, to reduce emissions to 80% of their 1990 levels by 2050. However, a gap exists between Obama's medium-term target and the expectations of the international community. At the climate change conference held in Copenhagen in 2009, the United States pledged to reduce its GHG emissions by 17%, in relation to its 2005 levels, by 2020. However, this is only 4% less than the 1990 level. There is a large gap between this commitment to emission reduction and that of the EU, which is 20%. Obama's commitment is also much lower than the recommended reduction of GHG emissions by 25%–40% (referring to 1990 levels) before 2020, proposed by the IPCC for industrialized countries in the developed world. Evidently, the commitment made by the United States regarding reduction of its emissions is far from being commensurate with its status as the world's major emitter. Therefore, it will be important to pay attention to the next round of interactions between the United States and other developed countries, especially those within the EU.

The changed attitude of the United States will, however, bring enormous pressure to bear on developing countries. When the United States is truly determined to make a change, the pressure will fall on large developing countries such as India and China. Developing countries, particularly large ones, must enter the "Post-Kyoto" era as soon as possible, following the entry of the United States.

China and the United States have long sparred with each other on the issue of the global climate. However, in reality, their attitude of "non-action" also provides a good excuse for the other party to refuse to reduce their emissions. China has long refused to accept a binding emission reduction index, in addition to adhering to the stance that it is a developing country. These standpoints also underlie its repeated criticism of the attitude of non-action on the part of the United States, which ranks highest, globally, for per capita emission. However, from the perspective of the current development trend, the increasingly positive attitude of the United States regarding the emission reduction issue has led to mounting pressure on China. A pressing question for Chinese scholars and the government, then, is how to participate in an international emission-reduction system in a way that is both consonant with domestic conditions and compliant with the global trend. With the issue of climate change having become a "politically correct" one, all countries can only adopt one of two standpoints: to oppose or to support emission reduction. These standpoints necessarily correspond to two consequences: supporters are admitted into the fold at moral commanding heights, while opponents become the target of public criticism and face consequences, notably a damaged international image as well as various trade sanctions.

It is worth reflecting on and learning from the strategy adopted by the United States, in which the aim is to stimulate the economy and create jobs while permanently solving the issue of energy efficiency through environmental policies and an energy revolution. China and the United States are in very similar situations in this regard. Both nations are strongly dependent on external energy sources, and both face an economic crisis and have simultaneously introduced massive economic stimulus packages and budgets. It remains to be seen whether China can take advantage of this opportunity to stimulate industrial restructuring, encourage the promotion of new energy-saving technologies, substantially increase its energy efficiency, and even stimulate its economy through green industry, becoming the first nation to emerge from the economic downturn. All of this will depend on what countermeasures China develops and the actions that it takes.

China and the United States are the largest developing country and developed country, respectively. They are also the two largest GHG emitters globally, with the annual GHG emissions of these two countries accounting for 42% of total global emissions. Therefore, they play an important role in addressing global climate change. Only through Sino-US cooperation can we achieve a global agreement to effectively address the challenge of climate change. Without this cooperation, it will be impossible for global efforts to successfully tackle climate change. Although the two countries differ in many respects, given variations in their energy sources, climates, and domestic conditions, they should "think big and start small" in seeking common ground while reserving the right to differ and engage in win-win cooperation on climate change issues. In the report *Overcoming Obstacles to U.S.-China Cooperation on Climate Change* (Meeting the Challenges of Clean Energy and Climate Change: Pathways to US-China Cooperation), Kenneth Lieberthal and David Sandalow have highlighted the fact that China and the United States should acknowledge the legitimacy of each other's perspectives and seek common ground on the nature of future commitment. The report further states:

> When it comes to cooperation on climate change and clean energy, the United States and China should think big and aim high. They are two great nations addressing one of the great challenges of our time. Thus, China and the United States should encourage all parties to engage in action programs to reach the 2050 global GHG emission reduction targets. They should treat climate change issues as pivotal for securing not only their national interests but also those of humanity as a whole.

Notes

1 Tony Blair, "Breaking the Climate Deadlock: A Global Deal for Our Low-Carbon Future," *Report submitted to the G8 Hokkaido Toyako Summit* (2008): 28.
2 International Monetary Fund, *World Economic Outlook: Housing and the Business Cycle, April 2008*, p. 166. www.imf.org/external/pubs/ft/weo/2008/01/pdf/text.pdf.
3 William R. Cline, "Global Warming and Agriculture," *Finance and Development* 3 (2008): 27.
4 International Energy Agency, *World Energy Outlook 2014*.
5 Blair, "Breaking the Climate Deadlock," 29.
6 McKinsey, *A Cost Curve for Greenhouse Gas Reduction*, 2007, 9–10.
7 McKinsey, *China's Green Revolution: Prioritizing Technologies to Achieve Energy and Environmental Sustainability*, 2009, 11–12.
8 Inge Kaul, Isabelle Grunberg, and Marc Stern, "Defining Global Public Goods," in *Global Public Goods: International Cooperation in the 21st Century*, eds. Inge Kaul, Isabelle Grunberg, and Marc Stern (New York: Oxford University Press, 1999).
9 IPCC consists of three Working Groups. The first Working Group is responsible for assessing the climate and related changes from a scientific perspective. It reports available knowledge on climate change such as how this happens and at what speed. The second Working Group is responsible for assessing the extent of damage inflicted by climate change on the society, economy, and natural ecology; the negative and positive impacts of climate change; and methods for adapting to changes. Thus, it assesses the impacts of climate change on humans and the environment, and how these effects can be reduced. The third Working Group is responsible for assessing the possibility of limiting GHG emissions or mitigating climate change. Thus, it investigates how to stop or slow down human-induced climate change.
10 Nicholas Stern, *Revelation of Key Points of Global Climate Change Agreement on China*, transcript of a speech delivered at the School of Public Policy and Management, Tsinghua University, 2008 (not reviewed by the speaker).
11 Thomas Schelling, "Addressing Greenhouse Gases," in *Nobel Masters toward Energy and Environment-2007 Nobel Laureates Beijing Forum* (Beijing: Science Press, 2008), 178–179.
12 International Energy Agency, *CO2 Emissions from Fuel Combustion Highlights*, 2014.
13 International Energy Agency, *2014 Key World Energy Statistics*.
14 Eric Posner and Cass Sunstein, "Pay China to Cut Emissions," *The Financial Times*, (UK) (Chinese edition), August 9, 2007. In 1987, the United States vigorously promoted the signing of the Montreal Protocol, which imposed limits on emissions of ozone-depleting chemicals. The United States did this not out of altruism, but because President Ronald Reagan was firmly convinced that the resulting gains for the United States would far exceed its losses based on a cost-benefit analysis. Prohibiting emissions of ozone-depleting chemicals does not pose a burden for companies in the United States. However, developing countries strongly opposed this protocol. They requested and received substantial compensatory payments from rich countries. www.ftchinese.com/story/001013326.
15 Zhang Huanbo, "Climate Protection Policy Simulation Study: Macro Dynamic Economic Model Based on Multi-National Climate Protection." PhD diss., Institute of Policy and Management, Chinese Academy of Sciences, 2007.
16 Data from 2012 or for that available for the most recent year. HDI data are typically calculated two years later.
17 US Energy Information Administration, *Annual Energy Outlook 2015 with Projections to 2040*.

4 Climate change and China

Threats and challenges

As the country that has endured the most frequent natural disasters and the largest losses in the world, China is one of the worst-hit victims of global climate change. China's processes of economic development, industrialization, and urbanization, as well as the formation of international division of labor system among countries, have led to mounting pressure on China's energy sources and environment. There is an urgent need to break with traditional patterns of economic growth and energy consumption. On the one hand, the development of a low-carbon economy under the new global trade terms will bring about a resurgence of trade protectionism. On the other hand, energy-saving and clean energy technologies, two examples of the efforts promoted to evolve a low-carbon economy, will provide technological support for China's initiatives to save energy and reduce carbon. The redefining of responsibilities will also enable China to obtain funds from developed countries to reduce its emissions.

4.1 Natural disasters and losses worldwide: China as the worst-hit victim

The world has now entered an era in which natural disasters are common, with their frequency and severity continuing to rise. From 1901 to 1910, 82 natural disasters were recorded, whereas more than 4,000 natural disasters were recorded for the period from 2003 to 2012.[1]

The world's total population is over 7 billion, with nearly a third of this population being affected each year by natural disasters. This affected population is mainly concentrated in developing countries. Faced with the same natural disaster as a developing country, a developed country can quickly respond to minimize economic losses and ensure that the smallest possible range of the population is affected. However, in developing countries, the number of people affected by a disaster far exceeds that in developed countries. In 2013, the top 10 deadliest natural disasters occurred mainly in developing countries within Asia.

The main locations of the top 10 disasters that affected the largest numbers of people in 2013 were the Philippines, India, China, and other developing countries in Asia. In 2013 alone, 4 of the top 10 global disasters that affected the largest

Table 4.1 The top 10 natural disasters with the highest death tolls in 2013

Disaster type	Country	Death toll
Typhoon (Haiyan)	Philippines	7,354
Flood	India	6,054
Heat wave	United Kingdom	760
Heat wave	India	557
Earthquake	Pakistan	399
Heat wave	Japan	338
Flood	Pakistan	234
Flood	China	233
Earthquake	Philippines	230
Flood	Cambodia	200

Source: Annual Disaster Statistical Review (2013).

Table 4.2 Top 10 global disasters affecting the largest numbers of people in 2013

Disaster type	Country	Month	Number of people affected (millions)
Typhoon (Haiyan)	Philippines	November	16.1
Typhoon (Phailin)	India	October	13.2
Typhoon (Utor)	China	August	8.0
Drought	China	January – July	5.0
Flood	China	July	3.5
Flood	Thailand	September – October	3.5
Earthquake	Philippines	October	3.2
Flood	Philippines	July	3.1
Drought	Zimbabwe	September	2.2
Earthquake	China	April	2.2

Source: Annual Disaster Statistical Review (2013).

numbers of people occurred in China, with the cumulative number of affected people reaching up to 18.7 million. Commencing from 1950, the frequency of catastrophic and major disasters has been on the rise. Generally, these natural disasters have occurred during the 1980s, the 1990s, and the early 21st century, providing some indication of the climate-related anomalies that have occurred at the onset of the current century.

During the 10 years spanning 1990 to 1999, the death toll caused by natural disasters in China was 34,000, accounting for 6.6% of total global natural disaster–related mortalities. The cumulative number of people affected in China was 1.263 billion, accounting for 65.8% of the global total. According to the available data, presented in Table 4.3, China is one of the locations with the most frequent occurrences of natural disasters and the greatest losses worldwide. Environmental and climate-related issues are therefore, without doubt, the greatest challenges facing China.

Table 4.3 The proportion of China's losses resulting from natural disasters to global losses (1979–2014)

	Number of disasters	Number of deaths (10,000)	Number of people affected (10,000)	Economic loss (US$100 million)
China				
1979–1989	98	1.78	0.98	94.96
1990–1999	175	3.40	12.63	1,178.86
1999–2008	252	1.13	10.48	750.49
2009–2014	310	1.52	5.65	1,280.76
World				
1979–1989	1,766	78.03	11.30	1,671.10
1990–1999	2,746	46.30	19.19	6,415.64
1999–2008	4,112	65.99	21.34	7,575.90
2009–2014	3,530	47.01	9.89	9,391.24
Proportion of China's disasters to that of the world (%)				
1979–1989	5.5	2.3	8.7	5.7
1990–1999	6.4	6.6	65.8	18.4
1999–2008	6.1	1.7	49.1	9.9
2009–2014	8.8	3.2	57.2	13.6

Source: Authors' calculations based on data compiled from the International Disaster Database of the Centre for Renewable Energy Development (CRED).

Note: Natural disasters included in the statistics meet at least one of the following conditions: (1) at least 10 people are killed, (2) at least 100 people are affected, (3) a state of emergency has been declared, or (4) international assistance has been requested.

China's history is one replete with natural disasters. Over the course of a 2,000-year time span, from the Zhou Dynasty to the Qing Dynasty, China suffered 1,052 droughts, 1,029 floods, and 473 locust plagues (see Table 4.3). China's overall disaster situation has worsened over the years and demonstrates a high level of volatility. An analysis of disaster situations in relation to demographic data indicates population declines during periods with higher occurrences of natural disasters. This is because population growth results in increased reclamation of cultivated lands and enclosure of lakes, rivers, or even tideland for cultivation, causing damage to the natural environment. Humans thus become the victims of self-inflicted disasters (see Table 4.4).

Let us consider, as an example, the occurrence of floods and droughts as indicators of natural climate change that reflect human destruction of nature from the time of modern China's establishment. As Table 4.5 shows, from 1950 to 2013, the number of disaster-affected and damaged areas in China increased. A disaster-affected area is defined as one that is affected but not necessarily damaged by a disaster. A damaged area is defined as one in which production is reduced by up to 30% below the usual annual production amount as a result of the disaster. During the 1950s, the average area annually affected by floods in China was 7,891,300

Table 4.4 Statistics on China's disaster history (Zhou Dynasty to Qing Dynasty)

Dynasty	Number of calculated constant years	Droughts	Floods	Total number of disasters	Average number of droughts per annum	Average number of floods per annum	Average number of disasters per annum
Zhou	867	30	16	89	0.035	0.018	0.103
Qin and Han Dynasties	440	81	76	375	0.184	0.173	0.852
Wei and Jin Dynasties	200	60	56	304	0.300	0.280	1.520
Northern and Southern Dynasties	169	77	77	315	0.456	0.456	1.864
Sui Dynasty	29	9	5	22	0.310	0.172	0.759
Tang Dynasty	289	125	115	493	0.433	0.398	1.706
Five Dynasties	54	26	11	51	0.481	0.204	0.944
Song Dynasty	319	183	193	874	0.574	0.605	2.740
Yuan Dynasty	97	86	92	513	0.887	0.948	5.289
Ming Dynasty	267	174	196	1,011	0.652	0.734	3.787
Qing Dynasty	268	201	192	1,121	0.750	0.716	4.183
Total	2,999	1,052	1,029	5,168	0.351	0.343	1.723

Source: Xiao Guoliang, *Imperial Power and Chinese Society and Economy* (Xinhua Publishing House, 1991).

Table 4.5 Areas affected and damaged by floods and droughts (1950–2013)

Years	Flood			Drought		
	Affected area (10,000 hectares/ year)	Damaged area (10,000 hectares/ year)	Damage rate (%)	Affected area (10,000 hectares/ year)	Damaged area (10,000 hectares/ year)	Damage rate (%)
1950–1959	789.13	496.25	57.53	1,322.38	416.63	34.11
1960–1966	942.00	585.43	57.74	2,164.71	1,002.57	45.80
1970–1979	535.70	224.30	39.64	2,164.10	750.00	28.02
1980–1989	1,042.50	552.90	52.71	2,463.80	1,176.10	47.56
1990–2000	1,459.36	923.00	63.20	2,632.27	1,331.82	50.60
2001–2011	11,272.70	5,943.90	52.73	23,981.20	13,346.40	55.65
2012–2013	2,264.73	292.42	12.91	2,344.02	179.01	7.64

Note: Agricultural disaster indicators were adjusted in the *China Statistical Yearbook* from 2012 onward. Consequently, "floods" that occurred in 2012 and 2013, shown in the table, include disasters caused by floods, landslides, mudslides, and typhoons, and the damaged areas also contained areas unable to harvest due to natural disasters.

Source: Author's calculation based on data derived from the *China Statistical Yearbooks*.

Table 4.6 Nationwide reductions in average annual grain outputs caused by natural disasters (1952–2013)

Years	Reductions in annual grain outputs caused by natural disasters (10,000 tons)	Annual grain outputs (10,000 tons)	Proportion of losses to total outputs (%)
1952–1959	379.49 (151.79)	18,025.125 (1,437.224)	2.1 (0.8)
1960–1966	612.26 (170.73)	17,386.143 (2,597.466)	3.50 (1.00)
1970–1979	662.72 (356)	27,612.400 (2,896.999)	2.40 (1.30)
1980–1989	1,595.12 (325.02)	27,699 (3,254.569)	4.20 (0.90)
1990–2000	3,290.89 (641.60)	47,036.030 (2,903.380)	7.00 (1.37)
2001–2011	3,571.17 (806.87)	49,734.010 (4,360.200)	7.30 (2.13)
2012–2013	499.43 (255.07)	59,575.900 (873.890)	0.80 (0.40)

Note: The figures in parentheses denote standard deviations, and the grain losses have been calculated as a proportion of 30% of the output during normal years. Agricultural disaster indicators were adjusted in the *China Statistical Yearbook* after 2012, so the damaged areas in 2012 and 2013 became unable to harvest.

Source: Author's compilation and calculation based on data extracted from *China Statistical Yearbooks*.

hectares. In the 1990s, this area had increased to 14,593,600 hectares per annum, and is currently measured at 10,413,000 hectares per annum. Drought-affected areas have also been expanding every year, and currently cover more than 23 million hectares.

Table 4.6 shows calculations of reductions in China's average annual grain outputs in areas damaged by natural disasters.

Table 4.6 shows that in the 1950s, the reduction in grain outputs caused by natural disasters was 3.7949 million tons, amounting to 2.1% of total grain outputs (the standard deviation was 0.8). However, from 1993 to 1998, the reduction in grain outputs caused by natural disasters increased sharply. This is because with the continued increase in monocrop production, the same type and magnitude of natural disaster that previously occurred will lead to higher economic losses. Thus, in the face of accelerated natural disasters and improved land productivity, disaster prevention and reduction means increased food output. While China is the largest producer of cereals and grains in the world, it is also the country with the greatest output reductions (losses) attributed to natural disasters. Natural disasters and abnormal weather patterns caused by global climate change pose the greatest threat to Chinese agriculture and farmers. Consequently, research on disaster prevention and mitigation is of great significance.

Table 4.7 Nationwide economic losses directly attributed to disasters (1998–2013)

Year	Economic losses calculated at costs for specified years (Unit: RMB 100 million)	GDP calculated at costs for specified years (Unit: RMB 100 million)	Proportion to GDP (%)
1998	3,007	83,024.3	3.6
1999	1,962	88,479.2	5.6
2000	2,045	98,000.5	3.2
2001	1,942	108,068.2	2.8
2002	1,717	119,095.7	3.9
2003	1,884	134,977.0	3.6
2004	1,602	159,453.6	2.2
2005	2,042	183,617.4	2.1
2006	2,528	215,904.4	1.8
2007	2,363	266,422.0	1.4
2008	11,752	316,030.3	1.4
2009	2,523	340,320.0	1.0
2010	5,340	399,759.5	1.1
2011	3,096	468,562.4	1.2
2012	4,186	518,214.7	0.9
2013	5,808	566,130.2	3.7

Source: Authors' calculation based on data derived from the *China Statistical Yearbook 2014*, as well as the *Social Service Development Statistical Bulletin 2010–2013* and the *Civil Affairs Development Statistical Bulletin 1999–2009*, issued by the Ministry of Civil Affairs.

The proportion of natural disaster–related economic losses to the GDP and newly added GDP is very high. Therefore, disaster reduction implies an increase in GDP (Table 4.7).

4.2 China as the worst-hit victim of global warming

The IPCC report has observed that the Himalayan glaciers are shrinking as a result of global warming. Moreover, it notes that the continued temperature rise will eventually result in a northward movement of China's temperate zone and an expansion of the arid area. Cities like Shanghai will face increasingly frequent and serious heat waves that will cause considerable discomfort for their rapidly expanding populations. The UNDP report further notes that at its current pace, global warming will lead to the disappearance of two-thirds of the glaciers within China's territory, including Mount Tianshan, by 2060 and a total melting of the remaining glaciers by 2100. Serving as a barometer of global climate conditions and also marking the origin point of the Yellow and Yangtze Rivers, the Tibetan Plateau's glaciers are melting at an annual rate of 7%. In any climate change scenario, if the temperature rises above 2 °C (considered a threshold value of danger in relation to climate change), the rate of this glacier retreat will accelerate. As water resources stored within the glacial water bank gradually reach a point of exhaustion, the water flow will decrease, affecting seven major Asian river systems: the Brahmaputra, Ganges, Salween, Yellow, Indus, Mekong, and Yangtze. These river systems collectively provide water and maintain the food supplies of more than two billion people. The forecast also shows that by 2050, the flow of the Brahmaputra River will be reduced by 14%–20%.

Because China has relatively less precipitation in general, arid and semiarid areas, which account for half of the area of its national territory, are directly and seriously affected by anomalies of climate change (Table 4.8). These effects are more serious than those on areas of the same latitude in other countries. If we compare the natural conditions of Japan and China, the average annual precipitation on Japan's land area of 378,000 km² is almost over 1,000 mm, with most areas over 1,500 mm or more. By contrast, humid regions comprise only one-third of China's total land area, with annual precipitation in coastal regions in South China being in the range of 1,600 to 2,000 mm. Precipitation in the area of the

Table 4.8 Distribution of arid and humid areas in China

Area	Proportion to the area of the national territory (%)	Geographical distribution	Arid and humid conditions	Vegetation	Agriculture and animal husbandry
Humid area	32	To the south of the Qinling Mountains and Huaihe River and in the eastern part of northeastern China	Annual precipitation > 800 mm	Forest	Paddy fields
Semi-humid area	15	Northeastern China and the North China Plain, the southeastern part of the Loess Plateau, and Qinghai-Tibet Plateau	Annual precipitation > 400 mm Precipitation > evaporation	Forest Grassland	Dry land
Semiarid area	22	Most parts of Inner Mongolia and the Loess and Tibetan Plateaus	Annual precipitation < 400 mm Precipitation < evaporation	Grassland	Irrigated agriculture Animal husbandry
Arid area	31	The western parts of Xinjiang, and the Inner Mongolian Plateau and the northwestern part of the Qinghai-Tibet Plateau	Annual precipitation < 200 mm Precipitation < evaporation	Desert	Oasis agriculture Animal husbandry

Source: *China Statistical Yearbook 2011.*

Yangtze River and regions to its south is over 1,000 mm, whereas in northern and northeastern China, it is only 400–800 mm. Inland northwestern regions experience 100 to 200 mm of precipitation, while this is less than 25 mm in the Tarim, Turpan, and Qaidam Basins.[2] A comparison of temperature rises in Japan and China shows that the extent and degree of damage faced by China exceeds that confronting Japan. From the perspective of our own fundamental interests and objective of long-term sustainable development, China is willing to take a more positive political attitude and implement more initiatives responding to our fundamental interests to meet long-term sustainable development objectives and join in the global response to global warming.

The northern, western, and eastern regions of China will be severely affected by climate change. The UNDP's report points out that the northern part of China has become one of the areas experiencing the most serious water shortage, worldwide. In some areas of the "3-H" (Haihe, Huaihe, and Yellow River) basins, the current level of exploitation of renewable water resources is 140%. This explains the rapid shrinkage of the main water systems and the decline in groundwater levels. In the medium term, the change in the melting pattern of glaciers will exacerbate water shortages. About half of China's 128 million rural poor live in this region, which accounts for 40% of China's total agricultural land and one-third of its GDP through the value of its outputs. Consequently, glacial melting has a serious impact on China's development. The entire ecosystem of western China is under threat. By 2050, it is predicted that the region's temperature will increase by 1–2.5 °C. The area of the Qinghai-Tibet Plateau is equivalent to the entire area of Western Europe, and there are more than 45,000 glaciers located in this region. However, these glaciers have been shrinking at an alarming annual rate of 131.4 km[2]. A report titled *China's National Climate Change Programme* also points out that the glaciers in northwestern China have been reduced by 21% in approximately the last 50 years. Moreover, permafrost in Tibet has undergone thinning of up to 4–5 m.[3]

The eastern region, represented by Shanghai, is particularly vulnerable to climate-related disaster hazards. Located in the estuary of the Yangtze River, Shanghai is only 4 m above sea level, and thus faces serious flood risks. Summer typhoons, storm surges, and river rises can cause major floods at any time that threaten 24.15 million Shanghai residents. Because of rising sea levels, there are increasing storm surges, and the coastal city has been placed on the danger list. However, the most vulnerable people are the 2.51 million temporary residents of Shanghai who are from rural areas. Most of these people live in shacks located near construction sites or flood-prone areas. They are and always will be vulnerable to enormous hazards because of their limited rights.

China's National Climate Change Programme summarizes overall national conditions relating to global climate change as follows: poor weather conditions, severe natural disasters, a fragile ecological environment, a coal-dominated energy structure, a large population, and a low level of economic development.[4] As the country with the world's largest population, and given its vast territory and complex natural ecosystems, China will be one of the countries hardest hit by global climate change. For example, with regard to the frequent meteorological disasters that affect China, the extent of the affected areas, the multiple types of disasters, the degree of their severity, and the size of affected populations are rarely comparable in other parts of the world.[5] The impacts of climate change on China are reflected in its agriculture, forests, water resources, coastal regions, and other ecosystems. Climate change does, however, have some positive influences on agricultural production. For example, it helps to extend the growing period of some crops and to shorten the frost period. However, in general, the advantages that it offers to China's agricultural production are far outweighed by the severity of its disadvantages, and its evident impacts in other areas. The main observations on China's climate change, described in *China's National Climate Change Programme,* emphasize the fact that global climate change has already had an enormous impact on China and that this impact is likely to further intensify in the future.[6]

4.3   Resource and environment challenges for the long-term development of China

China's rise brings not only positive but also negative externalities to the world. The positive externalities refer to the scale effects of a huge population and labor participation, as well as a rapidly expanding economic scale effect in relation to open markets and trade. The negative externalities refer to the scale effect of being the largest resource consumer. China has become the world's factory, and countries worldwide are benefiting from the huge scale effect of China's economic rise. During the period from 1980 to 2013, the proportion of China's GDP, trade volume, and FDI to the total global volume rose sharply. As Table 4.9 shows, China produced 48.5% of the world's steel and 58.6% of its cement in 2013.

A longer-term estimate indicates that China will become a superpower, in the true sense, by 2030, and almost all of its main indicators will rank first in the world (see Table 4.9). The model for the rise of China as a "giant country" has not previously been seen. Ultra-large-scale energy resource consumption and ultra-large-scale GHG emissions are inevitable by-products of China's development

Table 4.9 China's significance within the global economy in proportion to total global volumes (1980–2013)

Unit: %

Year	1980	1995	2000	2005	2013
GDP (PPP in US dollars in 2006)	3.2	9.1	11.3	14.5	12.3
GDP (market exchange rate)	2.9	2.5	3.8	5.0	15.8
Trade	0.9	2.7	3.6	6.7	12.0
Foreign direct investment (FDI)	0.1	11.0	3.0	8.0	8.5
Steel production	8.2	13.0	15.5	31.2	48.5
Cement production	9.0	33.6	37.4	46.6	58.6

Sources: National Bureau of Statistics, World Bank WDI database, and UNCTAD.

Table 4.10 Proportion of China's main indicators to the total global volume (1950–2030)
Unit: %

Year	1950	1960	1978	1990	2005	2030
Total population[1]	21.65	21.94	22.30	21.6	20.50 (2003)	< 20.00[8]
Urban population	8.50[2]	13.05[2]	11.18[2] (1980)	13.42[2]	16.15[2] (2000)	18.34
GDP (ppp)	4.59	5.24	4.90	7.83	15.10 (2003)	23.10
Exports	0.60[3] (1948)	2.00[3]	0.80[3]	1.80[4]	8.20[3] (2006)	> 15.00[8]
Crude steel production	0.36	3.17	4.43	8.50	30.90	> 40.00[8]
Generating capacity[4]	0.48	2.58	3.34	5.27	12.50[7] (2004)	> 20.00[8]
Energy consumption		7.18[5] (1971)	8.51[5]	10.06[5]	15.24[6]	21.55[6]
CO_2 emission		8.26[5]	7.93[5]	10.66[5]	19.16[6]	27.32[6]

Sources:

[1] Angus Maddison, *Chinese Economic Performance in the Long Run: 960–2030 AD*, trans. Wu Xiaoying and Ma Debin, proofread by Wang Xiaolu (Shanghai: Shanghai People's Publishing House, 2008).

[2] Data for China have been extracted from the *China Statistical Yearbooks*, and global data have been obtained from the UN Department of Economic and Social Affairs, *World Urbanization Prospects* (revised 2003 edition).

[3] WTO: International Trade Statistics 2007, www.wto.org.

[4] Tsuneta Yano Commemoration: "Data Analysis of Japan 100 Years of Development," 2004.

[5] World Bank, *World Development Indicators 2007*, CD-ROM.

[6] IEA: *World Energy Outlook 2007*, www.iea.org.

[7] Tsuneta Yano Commemoration "Japan's Parliament," 2007.

[8] Estimate calculated by Hu Angang.

process. By 2030, China's energy consumption will account for slightly more than one-fifth of the world's total energy consumption, and CO_2 emissions will account for slightly more than a quarter of the world's total emissions.

Accelerated industrialization and urbanization have resulted in mounting pressure on China's resources, energy, and environment. At present, China is undergoing a stage of accelerated industrialization. Its average annual growth rate of industrial added value was 23.1% during the period from 2000 to 2011, reflecting an almost ninefold increase.[7] The development of the heavy chemical industry has accelerated from 1992 onward, and in 2012, the ratio of light and heavy industries was less than 3:7 (see Figure 4.1). China's industrial development model is still a typical "black" one. Its highest costs relate to energy consumption and pollutant emissions. From 2005 onward, the energy-intensive industries have grown rapidly, but overcapacity has been widespread because of the outbreak of the international financial crisis. Consequently, this trend has slightly eased. During the period from 2011 to 2013, raw iron, steel, and cement production increased by

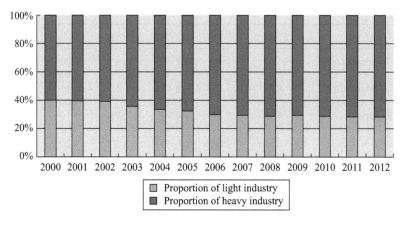

Figure 4.1 Proportions of heavy and light industries in China's industrialization process (2000–2012)

Note: Data on gross industrial outputs for light and heavy industries are shown for the period from 2000 to 2011, while the data for 2012 concerned the prime operating revenue generated from light and heavy industries. From the second half of 2013 onward, the categories of "light" and "heavy" industries have been replaced by the mining industry, the manufacturing industry, and the electricity, heat, gas, and water production and supply industry in the industrial data released by the National Bureau of Statistics.

Source: *China Statistical Yearbooks* of the National Bureau of Statistics.

5.2%, 6.6%, and 7.3%, respectively, indicating decreases of 70.5%, 66.4%, and 65.7%, respectively, from their production during the period of the 11th Five Year Plan. However, China's industrial energy consumption in 2013 was 2.525 billion tons of standard coal, accounting for about 70% of the total energy end-use consumption, which far exceeded the proportion of coal consumption in 1990 (36%). This quantity was also much higher than the average proportion of 14% for developed economies (OECD countries and regions).[8] From 1978 onward, there has been an ongoing rise in both the proportion of Chinese industrial outputs to total global outputs as well as the proportion of Chinese exports to global exports. China is currently the world's largest exporter and the second-largest importer. China's exports have surpassed those of Japan, the United States, and Germany, while the proportions of exports of these three countries to total global exports have declined over the same period.

China's increasing level of urbanization has also intensified the energy supply and demand. From 1978 to 2013, China's permanent urban population increased from 170 million to 730 million, and the urbanization rate increased from 17.9% to 53.7%. At present, China is undergoing a process of "catching up" urbanization. Based on the development goals proposed in the National New-Type Urbanization Plan (2014–2020), the urbanization rate of China's permanent population will reach 60% by 2020, and the urbanization rate of the registered population by households will reach 45%. Feng Fei (2007) has pointed out that although the industrial structure and the development paths of various countries during their rapid economic growth periods have differed, a common factor has been the convergence of the consumption structure. In other words, when per capita disposable incomes increase to a certain extent, the consumption structure of residents, especially that of urban residents, will inevitably be modified for items ranging from food, clothing, and daily supplies to housing and transport. Changes in the consumption structure bring about an alteration of the industrial structure with the former determining the latter. For example, during the period of the 10th Five Year Plan, China's urban per capita housing area increased by 28.8%, and the number of cars owned per 100 urban households increased 5.7 times. At present, China is undergoing a period of rapidly changing urbanization, which is increasing at an annual rate of 1.4%. According to national statistics, urban residents' per capita commercial energy consumption exceeds that of rural residents by 3.5 times. This factor, combined with substantial consumption of resources during the process of constructing urban infrastructure, is exacerbating the conflict between the supply and demand of resources in China.[9]

According to the traditional model of economic development, China's future growth in terms of per capita energy consumption is inevitable, and its per capita GDP and growth trends for per capita energy consumption are consistent. Moreover, because of heavy industrialization of the Chinese economy, per capita energy consumption has continued to grow rapidly in recent years (see Figure 4.2). From

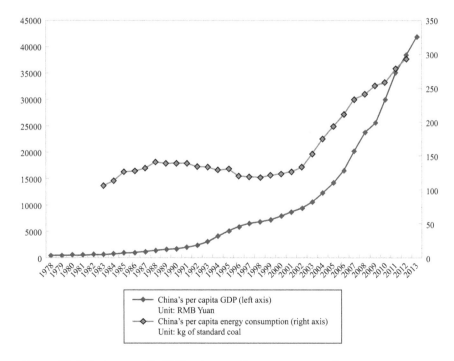

Figure 4.2 China's per capita GDP and per capita energy consumption

Source: *China Statistical Yearbook 2014.*

Table 4.11 International comparison of per capita primary energy consumption (oil equivalent in tons) for the period from 1820 to 2012

Year	World	UK	US	Japan	China
1820	0.21	0.61	2.45	0.20	
1870	0.31	2.21	2.45	0.20	
1913	0.61	3.24	4.47	0.42	
1950	0.84	3.14	5.68	0.54	
1973	1.54	3.93	8.19	2.98	0.48
1998	1.65	3.89	8.15	4.04	0.88
2005	1.78	3.88	7.89	4.50	1.32
2012	1.90	3.02	6.81	3.55	2.14

Sources: Tables 4 and 5 in Maddison, *The World Economy: Historical Statistics.* Data for the year 2012 were derived from the International Energy Agency, *Key World Energy Statistics* 2014.

the perspective of a global history of economic development, when a country's economy takes off, per capita energy consumption will show a high growth rate. An example is the economic takeoff of the United States during the period from 1870 to 1973 and Japan's economic takeoff during the period from 1950 to 1973. China's economic takeoff began after 1978, and by 2012, its per capita primary energy consumption was 2.14 tons of oil equivalent, which was even lower than the per capita consumption levels of the United States and Britain in 1870 (see Table 4.11).

In 2011, China became the world's largest energy consumer and the second-largest oil consumer after the United States. According to a report released by the US Energy Information Administration (EIA), China surpassed the United States as the world's largest net importer of oil by the end of 2013, and contributed 43% of the growth in global oil consumption in 2014. Moreover, China is currently the world's largest coal producer, consumer, and importer, with its coal consumption accounting for nearly half of the world's.[10] In 2013, the rate of China's primary energy consumption growth fell from 7% to 4.7%. Although this rate was considerably below that during the previous decade (8.6% per annum), it was still higher than the average global growth rate of primary energy consumption (2.3%).[11] It is estimated that based on current energy consumption patterns, China's energy consumption will continue to grow in the absence of relevant policies and measures. According to the Reference Scenario, China's average annual primary energy demand increased by 5.2% from 2005 to 2015, and the average annual projected growth rate was 3.2% for the period of 2005 to 2030. In *World Energy Outlook 2007,* IEA developed two scenarios for forecasting the growth of China's primary energy consumption for the period of 2005 to 2030 (see Table 4.12). The first alternative policy scenario entails the introduction of certain policies and measures to control energy consumption (this scenario is indicated by the superscript ① in Table 4.12). The second scenario (indicated by the superscript ② in Table 4.12) projects energy consumption under conditions of rapid economic growth (the average annual economic growth rate projected for 2005 to

Table 4.12 China's primary energy consumption forecast (2005–2030)
(Oil equivalent: one million tons)

	2005	2015①	2015②	2030①	2030②	Average growth rate from 2005 to 2030① (%)	Average growth rate from 2005 to 2030② (%)
Coal	1,094	1,766	2,075	1,861	3,003	2.1	4.1
Oil	327	517	625	652	1,048	2.8	4.8
Natural gas	42	126	125	225	284	6.9	7.9
Nuclear energy	14	44	34	120	82	9.0	7.4
Hydropower	34	70	63	99	91	4.4	4.0
Biomass and waste	224	218	229	251	219	0.5	–0.1
Other renewable energy sources		13	11	50	38	0.0	–
Total	1,735	2,754	3,163	3,257	4,765	2.6	4.1

Source: IEA, *World Energy Outlook, 2007.*
① The first scenario is described in the text.
② The second scenario is described in the text.

Table 4.13 Proportion (%) of primary energy consumption of the top four economies to total global energy consumption for the period from 1965 to 2035

	1965	1970	1980	1990	2000	2012	2035
United States	34.28	32.91	27.35	24.22	24.90	17.70	13
China	4.72	4.65	6.28	8.43	10.41	21.92	26
25 EU states	24.95	24.58	22.18	19.19	17.82	13.41	6
Former Soviet Union	16.19	15.48	17.36	17.54	10.14	8.24	5

Source: BP Statistical Review 2013, BP Energy Outlook 2035, Country and regional insights.
www.bp.com/en/global/corporate/about-bp/energy-economics/energy-outlook/country-and-regional-insights.html.

2030 is 7.5%). During this period, China's energy consumption growth rate will be maintained at 2.6%–4.1%, and the growth rate will not fall significantly.

Because of the large scale of China's total population, the proportion of the country's total energy consumption to that of the world is on the rise. It is estimated that this proportion will rise from 22% in 2012 to 26% in 2035 (see Table 4.13).

Growing dependence on foreign oil imports poses a serious threat to China's economic security. Because China became a net oil importer in 1993, its dependence on foreign oil imports has increased every year, reaching nearly 60% in 2013. It is predicted that China's dependence on oil imports will rise from about 60% (about 6.6 million barrels/day) in 2013 to 75% (13 million barrels/day) by 2035. This value is higher than the peak value of the United States in 2005. China's dependence on natural gas imports will simultaneously increase from the current level, which is slightly lower than 30% (4 billion cubic feet/day), to more than 40% (24 billion cubic feet/day).[12] The ratio of net oil imports of China's coastal regions may exceed 90%, and the gap between supply and demand will increase. The major oil-producing countries are those located in the Middle East and Africa, as well as Russia and Venezuela. They all either have the potential to become politically unstable, or typically use energy as a diplomatic weapon. Therefore, China's growing dependence on foreign oil imports from these oil-producing countries increases the uncertainty of its economic development and reduces room for maneuver for its foreign policy. In recent years, the global energy structure has been undergoing major changes. A rise in oil prices has changed the balance of power between oil-producing and oil-consuming countries, and while the traditional relationship of interdependence is disintegrating, a new order has not yet emerged. The Middle East will account for a growing proportion of the global oil supply, with more oil coming from unstable regions. This will consequently increase China's security costs. Moreover, the interaction between China and these countries will result in more suspicion and hostility between China and the United States, which will worsen their relations to some extent, heightening the urgency of security issues, including energy security, in a "self-fulfilling prophecy." Yeomans (2005) has observed that China's growing demand for oil means that the competition between the two countries for global oil resources will become increasingly fierce. A comprehensive exchange at the global level has been initiated between China and the United States on the issue of controlling the origin of oil.[13]

The comprehensive development of globalization and of the international division of labor has resulted in mounting pressure on China's resources and environment. The new pattern of global economic integration and specialization entails industry and export transfers, as well as energy consumption and

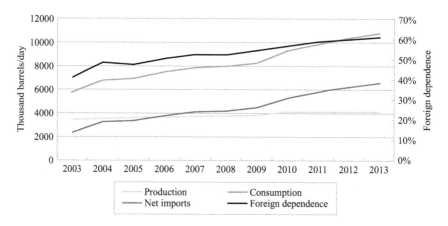

Figure 4.3 China's oil production, consumption, net imports, and foreign dependence (2003–2013)

Source: BP, Statistical Review of World Energy, 2014.

emission transfers. According to IEA's *World Energy Outlook 2007: China and India Insights,* the quantity of China's re-exported energy was 400 Mtoe in 2004, which was approximately 25% of China's total energy consumption for that year. Moreover, the energy content of China's imported goods was 171 Mtoe, which was equivalent to 10% of China's energy demand during that year. The proportion of energy in relation to China's exported goods is much higher than that of other countries. For example, in 2001, the proportions of re-exported energy of the United States, the EU, and Japan were 6%, 7%, and 10%, respectively. A high proportion of re-exported energy also causes an increase in CO_2 emissions. In 2004, the amount of CO_2 emissions associated with China's energy re-exports was 430 Mt, which was 26% of the carbon emissions of all sectors in 2001.

Studies conducted by the UK Tyndall Centre for Climate Change Research have also shown that in 2004, China's net exports caused the release of approximately 1.11 Gt of CO_2 emissions, accounting for 23% of China's total emissions (4.73 Gt) during that year. Thus, these emissions comprised almost a quarter of the total emissions in 2004. This amount was equivalent to Japan's total CO_2 emissions, and to the total combined amount of Germany's and Australia's emissions. It was more than double the UK's emissions for the same year.[14] Moreover, Liping, Yong, and Chunxiu (2008) have pointed out that only direct emissions were considered for the above studies, ignoring other investments that can result in significant emissions during the production process. Further, the carbon emission intensities of the largest export-contributing trade projects were slightly below the average carbon emission intensity. Therefore, while this figure also requires further verification using a more comprehensive input-output model, there should not be any significant difference.[15]

A considerable portion of the emissions resulting from China's energy re-exports can be attributed to energy consumption and carbon emissions from manufacturing industries that have been transferred from the United States, the EU, and Japan to China. A regional structural shift in global energy consumption and GHG emissions is also evident. As Table 4.14 shows, the proportions of emissions from the United States, the EU, and Japan have declined, while the proportions of emissions from large developing countries such as China and India have risen.

China's energy consumption structure is dominated by coal. The country's coal consumption initially increased and then decreased, remaining at a level of 65%–70% in recent years. Conversely, oil consumption initially decreased and then increased, remaining at a level of about 18% in recent years. Natural gas consumption, too, initially decreased and then increased, demonstrating a fast pace of growth. In 2013, China's natural gas consumption growth rate was 10.8%, ranking highest globally. In 2013, the growth of China's renewable energy sector, aimed at generating electricity (9.4 million tons of oil equivalent) and hydropower (8.9 million tons of oil equivalent), ranked highest globally, while the growth of nuclear energy (3 million tons of oil equivalent) was second only to that of the United States. Figure 4.4 shows the structure of China's energy consumption. The global structure of energy consumption shows a declining trend in relation to oil, coal, natural gas, hydropower, and nuclear energy. Globally, oil consumption

Table 4.14 The proportions of CO_2 emissions of six major economies in relation to total global CO_2 emissions (1960–2035)

Unit: %

	1960	1970	1980	1990	2012			2035^NPS	2035^450S
China	8.98	5.65	8.08	11.29	China	26.15	China	27.4	21.9
EU	15.87	15.09	13.59	10.96	OECD	38.71	OECD	27.3	27.3
United States	33.68	31.18	25.32	22.67	United States	16.17	Asia*	18.9	19.8
Japan	2.47	4.96	4.71	4.76	Japan	3.90	Middle East	6.4	7.5
Russia				9.26①	Russia	5.29	Non-OECD countries in Europe and Asia	8.0	9.3
India	1.28	1.30	1.79	3.01	India	6.23	Non-OECD countries in Americas	4.3	4.7
							Africa	3.7	3.6

Notes:

① Data in 1992.

NPS refers to the forecast under a new policies scenario, which is a scenario that follows the implementation of all of the policies currently being considered.

450S refers to the forecast under a "450 scenario." This is a scenario entailing long-term maximum greenhouse gas emissions (equivalent to a CO_2 level of 450 ppm) and efforts by the international community to control the global average temperature rise at up to 2 °C after 2013.

Sources: Data for 1960–1990: World Bank, World Development Indicator 2006, CD-ROM; 11 EU countries. Forecast data for 2012–2035: IEA, 2014 World Energy Statistics,* all Asian countries except China.

remains the dominant form of energy consumed, comprising about 39% of total energy consumption from 1990 onward. However, with soaring international oil prices in recent years, oil consumption has dropped significantly. Over the last decade, along with the rapid growth of hydropower and renewable energy consumption, the proportion of non-fossil fuels in electricity generation has gradually begun to rise. Renewable energy has become the main engine fueling the growth of non-fossil fuels, and the global share of renewable energy in electricity generation increased from 2.7% in 2008 to 5.3% in 2013. The impacts of Japan's Fukushima nuclear accident have reduced global nuclear energy consumption in recent years. This form of energy accounted for only 4.4% of the global energy consumption in 2013, which is close to the level of the 1980s.[16]

From 1978 to 2007, coal dominated China's energy consumption and at one stage accounted for 76% of the total energy consumed. The proportion of coal shows an overall downward trend, but in recent years, a sizable rebound has occurred, with coal consumption having increased dramatically (see Figure 4.5). The proportion of coal in the global energy consumption structure has been stable at 25%–29%. Rising oil and gas prices in recent years have been the main factors responsible for the rapid growth of coal consumption. According to the

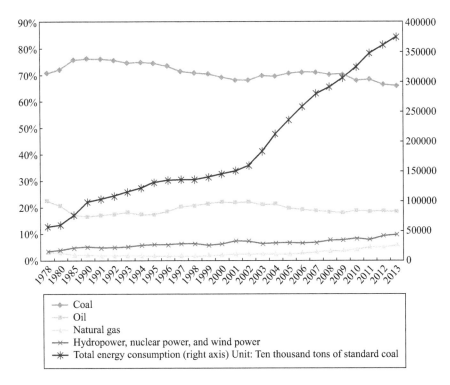

Figure 4.4 China's structure of energy consumption and total amounts of energy consumed (1978–2013)

Source: *China Statistical Yearbook 2014.*

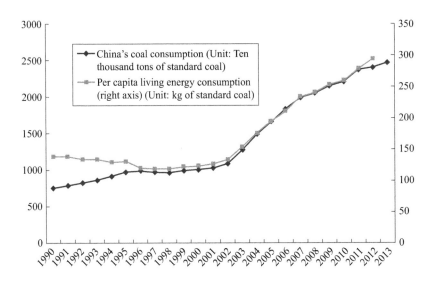

Figure 4.5 China's coal consumption (1990–2013)

Source: *China Statistical Yearbooks.*

data presented in BP's *World Energy Statistics 2013,* coal was still the fastest-growing fossil fuel in the world, accounting for 30% (the highest-recorded level since 1970) of the total global primary energy consumption that year. China is the world's largest coal producer and consumer. Its coal output in 2013 reached 1.84 billion tons of oil equivalent, reflecting an increase of 1.2% compared with the output in 2012, and accounting for 47.4% of global coal production, while coal consumption amounted to 1.93 billion tons of oil equivalent. This indicated an increase of 4.0% compared with coal consumption during the previous year. While the increase was the lowest since 2008, it still accounted for 67% of the global coal consumption increment, 50.3% of global coal consumption, and more than 70% of coal consumption in the Asia-Pacific region. The scale of coal utilization has brought huge external costs to China. Wastewater, gas, and residue are produced during coal mining, processing, transportation, and combustion processes, which have negative effects on the environment, ecology, and human health, all of which result in huge external environmental costs. External costs caused by coal mining include water and air pollution, noise pollution, impacts on aquifers, and overturning of the hydrological balance. Wastewater discharged from coal processing causes water pollution and destroys aquatic ecosystems. Coal transportation also has external costs, including residue dust. Air pollution caused by coal combustion, including emissions of dust, nitrogen oxides, carbon dioxide, and mercury, can cause asthma and other respiratory diseases, and can endanger the public's health, and even their lives. According to *Coal Environment External Cost Accounting and Internalization Scheme Research,* released by the Chinese Academy for Environmental Planning within the Ministry of Environmental Protection and Energy Foundation, China, the external environmental cost per ton of coal in China accounted for about 28% of the annual coal price. In 2010, China's total external environmental cost was RMB 555.54 billion, which was 2.3 times greater than the publicly financed cost of environmental protection.

In the context of rapidly rising oil prices and a high rate of economic growth, China's energy policy shows a noticeable swing. When environmental and climate issues become prominent, China's energy policy tends to shift to limit the proportion of coal within the energy structure, and to promote clean energy. When international oil prices rise, the policy supports the use of coal as a substitute for oil. During the recent period of soaring international oil prices, China's growth rate in coal consumption has significantly exceeded the growth rate of oil consumption. Long-term energy development and short-term economic development conflict with each other, and economic development priorities prevail at the expense of energy policy. Consequently, seesawing policies are inevitable.

In general, China's GDP energy consumption shows a downward trend (Figure 4.7). The cumulative energy saving rate of 64% was achieved within 20 years from 1980 to 2000. This was much higher than the rates of OECD countries and of the world, which were 19% and 20%, respectively. Starting in 2005, China's energy consumption per unit of GDP began to gradually decline, and significant progress was achieved in energy saving.

4.4 Challenges for China posed by a significant shift in global economic and trade patterns

In the international context of low-carbon economic development, China will be affected by a significant shift in global economic development and international trade patterns. Currently, focusing on energy and climate change policies, major developed countries such as the United Kingdom, the EU, Japan, the United States, and Australia have designed their own relative policies based on their individual socioeconomic contexts and have shifted toward a low-carbon future entailing different modalities.[17] International climate change negotiations currently focus on three main concerns: (1) how to reach a legally binding emission reduction agreement as soon as possible, (2) how to resolve the issue of INDCs in light of the controversy regarding the principle of "common but differentiated responsibilities" that is disputed by various parties, and (3) how to fulfill "green financing" and other commitments proposed at the Copenhagen Conference. Therefore, all countries had high expectations of the UNFCCC Paris Conference held in December 2015.

In this context, the traditional global pattern of trade will be replaced by a new one. Currently, green procurement public policies, designed to encourage trade in low-carbon products, are on the rise in Europe. Chatham Research Institute (2007) noted that although no formal mechanism existed to curb the trade of carbon-intensive industrial products, from a political point of view, it is easier to limit trade according to the impact of products (such as health or environmental impacts) than to limit trade according to the methods of producing the products. Many trade products (e.g., cement) are considered high-carbon products because of their production process, while others are categorized as being high-carbon because of their product characteristics and impacts (such as light bulbs and air conditioners). In the latter case, some product standards facilitate the trade of low-carbon products, effectively blocking the trade of carbon-intensive products (as in the case of building standards). In addition to formal standards, there are informal standards that play a similar role, such as the "Energy Star" rating of electrical equipment.[18]

EU countries have already begun to implement laws and policies to promote CO_2 emission reduction. However, they are worried about the loss of competitive advantages of their domestic products on the international market as a result of the increased energy costs entailed in emission reduction. Consequently, they have increased the threshold for foreign products to enter their countries through the use of green trade barriers (e.g., improving environmental standards and introducing a carbon tax). This implies that developing a low-carbon economy will result in a resurgence of trade protectionism, posing considerable challenges for the free trade model. Although China has maintained foreign trade surplus for many years, it has a resource-environment deficit in relation to foreign trade. China's trade structure entails the export of a large quantity of high energy-consuming and highly polluting products, with only a small quantity of low energy-consuming and low-polluting exports. Even considering similar products, the energy consumption and pollution intensity of exports per unit are much higher than those of similar products in developed countries. International experience shows that industrial transfer is an important means for achieving emission reductions. Many developed countries achieve significant emission reductions through industrial transfers and the adjustment of their domestic industrial structures. However, it will take a long time before China will be able to adjust its industrial structure and change its trade patterns.

Low-carbon technology innovation is at the core of the development of a low-carbon economy, and it determines whether the economy is energy saving and whether it reduces energy consumption and CO_2 emissions. It also underlies the development and use of renewable energy and optimizes the structure of energy consumption, which hinge on research and development as well as on the popularization and promotion of low-carbon technologies. The *National Outline for Medium and Long Term S&T Development* proposes specific targets for China's energy technology development. Thus, it is anticipated that by 2020, China will achieve a major breakthrough in energy development and energy-saving technologies, as well as clean energy technologies and an optimized energy structure. At the same time, China's main manufacturing industries are expected to reach or approach the energy efficiency levels of advanced countries. Innovation and development of low-carbon technologies will greatly enhance China's overall competitiveness. However, the low-carbon standards of developed countries have created major obstacles for Chinese exports. For example, EU member states have established a variety of standards aimed at protecting the public's health and the environment. Many of these standards are associated with climate change. For example, the emission reduction targets set by the EU according to the binding Kyoto Protocol have to some extent provided the impetus for the development of energy efficiency standards. These standards often entail different requirements for different products. The development of higher standards not only has a potential role in promoting trade, but also serves as a bargaining chip during bilateral negotiations, as lower tariffs will be charged for qualifying products.[19] China is faced with the task of developing unified standards for export destination countries or jointly developing standards.

The redefinition of carbon emission responsibilities also has implications for global trade. The Kyoto Protocol does not take into account consumers' responsibilities in relation to carbon emissions, holding that only producers are responsible for these emissions. Consequently, the conventional method of calculating carbon emissions is very biased against manufacturing powers such as China. At present, there is a mounting need to change the traditional method of calculating carbon emissions, as well as to redefine the parties responsible for carbon emissions. In the event the parties responsible for carbon emission are redefined, traditional trade policies (such as tariffs and non-tariff barriers) will also need to be redefined. The distribution of global supply chains in various countries is facing redeployment, and the global trade mechanism and pattern of economic development is also likely to face major changes. This prospect, on the one hand, would bring great opportunities for China, which would no longer have to bear the costs of carbon emission on behalf of consumers of exporting destination countries. Moreover, being a large manufacturing country, China could access the potential advantages of emission reduction to obtain a significant amount of emission reduction funds from developed countries. On the other hand, this would pose a major challenge for China, which would continue to be locked into an industrial chain characterized by high energy consumption, high pollution, and high emissions as a result of what is known as a "lock-in effect."

Notes

1 Even accounting for more comprehensive records being available for the latter period, this still represents a significant increase.
2 National Bureau of Statistics, *China Statistical Yearbook* (Beijing: China Statistics Press, 2007), 4.
3 The National Development and Reform Commission, *China's National Climate Change Programme,* 2007.
4 Ibid.
5 The National Development and Reform Commission, China's National Climate Change Programme, 2007.
6 The National Development and Reform Commission, China's National Climate Change Programme, 2007.
7 Source: *China Statistical Yearbooks* of the National Bureau of Statistics.
8 International Energy Agency, *World Energy Outlook 2014.*
9 Feng Fei, "Accelerate Resource-Saving Society Construction and Promote Economic Growth Mode Transformation," *Electric Power Technologic Economics* 19 (3) (2007): 1–5.
10 US Energy Information Administration (EIA), www.eia.gov/beta/international/analysis. cfm?iso=CHN.
11 BP, *Statistical Review of World Energy,* 2014.
12 *BP Energy Outlook 2035, Country Insights: China.* www.bp.com/en/global/corporate/ about-bp/energy-economics/energy-outlook/country-and-regional-insights/china-insights.html.
13 Matthew Yeomans, "Crude Politics: The United States, China, and the Race for Oil Security," *The Atlantic Monthly* 295 (4) (2005): 48–49.
14 Tao Wang and Jim Watson, "Who Owns China's Carbon Emissions?" *Tyndall Briefing Note No. 23,* October 2007, http://tyndall.webapp1.uea.ac.uk/publications/briefing_notes/bn23.pdf.
15 Li Liping, Ren Yong, and Tian Chunxiu, "Analysis of China's Carbon Emission Responsibility from the International Trade Perspective," *Environmental Protection* 3 (2008): 62–64.
16 Data is quoted from BP, *Statistical Review of World Energy 2014.*
17 Wang Yi, "Exploring a Low-Carbon Development Path with Chinese Characteristics," *Green Leaf* 8 (2008).
18 Chatham House (Royal Institute of International Affairs), *Interdependencies on Energy and Climate Security for China and Europe,* 2007, 57.
19 Ibid., p. 61.

5 National governance of climate change

A roadmap for China's emission reduction

China's current energy policy is based on holistic socioeconomic objectives of simultaneously achieving sustainable economic development and energy conservation. However, long-term strategies and viable targets relating to the implementation of a "Blueprints" policy are required for national climate change governance. While China has made clear commitments to reduce emissions, the next step will be to figure out how this can be achieved. This chapter proposes an emission reduction path response for China that is synchronous with the response at the global scale. This entails three steps, and it divides countries around the world into four categories according to varying levels of regional development and net carbon sources upon which emission reduction targets are based. To accomplish this goal, the first step entails a long-term investment in climate change. China evidently has the full capability and financial power required to make this investment.

5.1 Two scenarios for China's energy development: "Scramble" and "Blueprints"

In its report titled *Shell Energy Scenarios to 2050* (2008), Royal Dutch Shell presented two energy scenarios entailing contrasting development paths. The first, called the "Scramble," describes a situation in which policymakers pay little attention to making more efficient use of energy prior to experiencing a shortage in the energy supply, and do not seriously respond to the GHG problem before tangible effects of climate change materialize. In the second "Blueprints" scenario, people take increasing actions within their localities to deal with the challenges posed by economic development, energy security, and environmental pollution. This requires a policy that introduces payment of fees for GHG emissions. Such a policy would greatly promote the development of CO_2 capture and storage, as well as other clean energy technologies and measures to improve energy efficiency. Consequently, CO_2 emissions would be greatly reduced. Both of these scenarios pose challenges, because the world can no longer avoid the following hard truths about energy supply and demand.

- Changes in energy use are only occurring gradually.
- Supply is struggling to keep pace with demand.
- Environmental stresses are increasing.

As one of the world's largest energy consumers and GHG emitters, China is also facing two challenging and uncertain scenarios. The current reality of China's energy consumption fits the Scramble model. As discussed in the previous chapter, China's increasing dependence on foreign oil poses a serious threat to its national economic security, which not only increases the uncertainty of economic development, but also reduces the space for its foreign policy. However, it is difficult to change the coal-dominated energy consumption structure in the short term. During a period of soaring oil prices, coal consumption rises sharply. Continuous improvements in infrastructure required for ongoing industrialization and increasing urbanization levels have resulted in increasing pressure being put on China's resources and environment. Moreover, in this era of economic globalization, the international division of labor has resulted in a situation in which China needs to emit excessive GHGs for other countries to maintain its commodity exports. As previously stated, China's energy policy has been designed to accomplish its broader socioeconomic objectives. Consequently, achieving sustainable economic development and reducing energy density will be difficult to accomplish. China thus needs to develop both long-term strategies and viable objectives to avoid the Scramble and attain the Blueprints scenario.

China cannot avoid facing two of the realities described above: the tension between supply and demand and environmental stress. The Scramble scenario is not sustainable, as the continued supply and demand of declining fossil energy resources will inevitably lead to an unsustainable economy. The reason for the serious contradiction between supply and demand lies in energy consumption based on the current production model. Continued application of this model will inevitably create an unsustainable economic situation and the outbreak of a financial crisis similar to the 2008 global financial meltdown that originated in the subprime mortgage crisis.

If we combine the two energy realities confronting China and the two scenarios, we will reach a crossroads leading to diverging paths. Table 5.1 shows the differing outcomes of the two realities in relation to the two scenarios. On the road toward the Scramble scenario, the existing production and consumption chain determines that the contradiction between energy supply and demand will repeatedly surface and be intensified, along with increasing pressure placed on the environment. A contrasting route is that of sustainable development through the adoption of new patterns of energy coordination and use entailed in the Blueprints

Table 5.1 Two realities and two roads

	Supply and demand contradiction	*Environmental pressure*
Scramble	Non-sustainable economy	Environmental pollution and global warming
Blueprints	Sustainable economy	Environmental improvement and normal climate

scenario. Consequently, environmental stresses will be alleviated in a direction that enables human adaptation.

Evidently, the commitment to reduce emission provides China with an opportunity to attain the Blueprints situation, as well as to introduce domestic economic reforms and improve management systems. In the future, China will engage in international negotiations and global governance on climate change and will accept some climate change–related rules that can provide the impetus and opportunities for implementing sustainable energy policies and good environmental governance. More importantly, the grim climate change situation will exacerbate the pressure on countries to reduce emissions. To choose not to actively participate in global emission reduction efforts implies the choice to engage in a war over public resources – a black war without the smoke of gunpowder. China's selection of this path would result in its passive involvement in a worldwide Scramble situation.

5.2   Two constraints compelling China to undertake its emission reduction obligations

Currently, both the Kyoto Protocol and the Bali Roadmap reductively cleave the world's population of more than six billion and over 200 countries into two distinct groups: developing and developed countries. This division gives rise to the illusion among Chinese leaders that China will remain a developing country in perpetuity. However, China is not a typical developing country. Rather, it is on a course of continual advancement. The following two constraints determine that China must undertake emission reduction obligations.

The first constraint arises from the fact that the majority of the Chinese population is positioned at the high and upper-medium levels of human development. To some extent, China exhibits the characteristics of both developing and developed countries. More than two decades ago, the vast majority of China's population was positioned at the Third World and Fourth World levels. However, today almost half of China's population is situated within the Second World and the Third World. According to data for 2010, China is no longer a developing country in the typical sense. Beijing and Shanghai, which account for 3.20% of the total Chinese population, have reached an extremely high level of human development, advancing into the First World. Moreover, 42.01% of the country's total population has reached a high level of human development corresponding to the Second World. Overall, China's population has breached the Third World threshold, reaching a medium level of human development. Our analysis of 31 regions, nationwide, applying the HDI four-group classification method, shows that China already exhibits a pattern of one country and three worlds in relation to levels of human development. This implies that over a period of more than two decades, the regional pattern of human development in China has undergone a fundamental shift entailing a quantum leap from the Third World and Fourth World to the Second World and Third World. Over the course of 24 years, from 1982 to 2006, the proportion of China's First World population increased by 3.20%, that of its

Second World population increased by 36.50%, and that of its Third World population increased by 1.15%. Concurrently, China's Fourth World population has decreased by 41.36%. The change in this pattern reflects the advancement of China's overall level of human development, which is set to move to a higher level every other period (about a decade).

By 2020, China's HDI will reach 0.87–0.88, with the per capita GDP just reaching the global average, and China will belong to the group with a high level of human development as a whole. This position reflects the most important features of China's development model. Whereas the level of per capita income is still relatively low compared with the levels of the developed countries, the Chinese population has attained a high level of human development in terms of people's livelihood. At the same time, the HDI level in each region of China will demonstrate varying degrees of improvement. About 70% of the population will enter the level of the First World, and the Second World population will correspondingly decrease from more than 70% to more than 20%. In tandem, China will no longer have a Third World population. If China achieves this target smoothly, provinces with the lowest HDI levels globally, such as Tibet, are likely to enter the Second World, or at least the gap will be considerably reduced. From an international perspective, it is entirely possible for China to achieve the ambitious goal of building a moderately prosperous society by 2020. From the emission reduction perspective, China is more able to assume a global duty (Table 5.2).

Table 5.2 Proportion of populations of different HDI groups to the total population of China (1982–2006)

Unit: %

HDI group	1982	1990	2003	2006	Emission reduction conditions
First World	0.00	0.00	22.15	30.34	Unconditional emission reduction
Second world	2.10	37.45	74.63	69.44	Conditional emission reduction
Third World	56.54	62.36	3.22	0.21	Proposed emission reduction
Fourth World	41.36	0.19	0.00	0.00	Proposed emission reduction
Total	100.00	100.00	100.00	100.00	

Note: HDI refers to the Human Development Index. Population proportion refers to the proportion of the regional population in the same world to the total population of the country.

Data Sources: Population Census Office of the State Council & Department of Population Statistics, the National Bureau of Statistics, *China's 1982 Population Census* (digital collection) (Beijing: China Statistics Press, 1985); Population Census Office under the State Council, *Main Data of the Fourth National Census of China* (offline collection); National Bureau of Statistics, *China Population Statistics Yearbook 2004* (Beijing: China Statistics Press, 2004); Hu Angang and Zhang Ning, "Evolution of Regional Pattern and Disparities of China's Human Development (1982–2003)," in *National Report* (Institute for Contemporary China Studies Tsinghua University, 2006).

We can calculate the total net amount of carbon sources of regions in China based on carbon sources and the total carbon sinks of provinces (autonomous regions and municipalities).[1] The top 10 ranking Chinese provinces for net carbon sources are Hebei, Shandong, Shanxi, Henan, Jiangsu, Guangdong, Liaoning, Zhejiang, Hubei, and Hunan. The net carbon sources of these 10 provinces collectively account for 70.32% of China's total net carbon source. Consequently, these provinces should share emission reduction credits according to their negative externalities in relation to the country.

Beijing, Shanghai, Shandong, Zhejiang, Liaoning, Guangdong, Jiangsu, and Tianjin belong to the First World. The net carbon sources of Shanghai and Beijing account for only 4.27% of the total national net carbon source. However, these are First World regions with an extremely high level of human development. Moreover, they belong to the category of regions that should unconditionally reduce their emissions. Consequently, their emission reduction roadmaps should be synchronized with the global mission roadmap. The net carbon sources of Shandong, Zhejiang, Liaoning, Guangdong, Jiangsu, and Tianjin, which are Second World regions with a high level of human development, account for 33.19% of the total national net carbon source. Thus, these six provinces and cities should implement unconditional emission reduction, and their reduction rates should also be much higher than the one indicated for the global reduction roadmap (see Tables 5.3 and 5.4). Our initial premise was that these provinces, as regions demonstrating high levels of human development and high net carbon sources, required more aggressive reduction targets and should undergo the following stages in their emission reduction roadmaps:

Stage 1: Strive to reach peak carbon emissions before 2015.
Stage 2: Achieve a significant decline in carbon emissions after 2020 down to 1990 levels.
Stage 3: By 2030, reduce carbon emissions by half with reference to 1990 levels.

Hebei, Shanxi, Henan, Hubei, and Hunan belong to the second category. The collective net carbon sources of these five provinces, which are high-emission regions, account for 38.64% of the total national net carbon source. However, because there is a medium level of human development in these provinces, they also belong to the group of proposed emission reduction regions, and their emission reduction rates depend on the gap between their net carbon source levels and the national average level, as well as on the gap between the provincial HDI and the value of 0.8 (the lowest level of the First World).

Yunnan, Qinghai, and Tibet belong to the third category. The net carbon source of each of these three provinces is negative. In other words, they are negative-emission regions. The absolute value of their net carbon source accounts for 2.23% of the total net carbon source of other provinces, and they should thus be entitled to obtain ecological compensation from other provinces.

Table 5.3 Carbon sources, total carbon sinks, net carbon sources, and HDIs of various regions in China

Unit: ten thousand tons

Region	Total carbon source (2007)	Total carbon sink (2007)	Net carbon source (2007)	HDI (2010)
Hebei	85,057.81	1,146.25	83,911.56	0.691
Shandong	51,772.96	1,220.47	50,552.49	0.721
Shanxi	43,357.59	1,368.26	41,989.33	0.693
Henan	34,596.16	1,945.13	32,651.03	0.677
Jiangsu	33,044.54	524.34	32,520.20	0.748
Guang dong	29,906.58	2,819.13	27,087.45	0.730
Liaoning	25,409.13	2,229.92	23,179.21	0.740
Zhejiang	22,968.93	152.37	22,816.56	0.744
Hubei	18,299.45	802.57	17,496.88	0.696
Hunan	16,605.95	2,207.72	14,398.23	0.681
Anhui	14,770.68	808.69	13,961.99	0.660
Guizhou	15,788.25	2,056.16	13,732.09	0.598
Shanghai	13,466.64	18.89	13,447.75	0.814
Sichuan	17,666.72	4,478.25	13,188.47	0.662
Heilong jiang	16,642.27	3,780.58	12,861.69	0.704
Shaanxi	13,215.00	1,364.07	11,850.93	0.695
Inner Mongolia	26,590.98	15,212.24	11,378.74	0.722
Jilin	13,117.70	3,554.43	9,563.27	0.715
Fujian	10,852.59	2,564.52	8,288.07	0.714
Beijing	7,694.30	93.23	7,601.07	0.821
Tianjin	7,444.78	17.23	7,427.55	0.795
Xinjiang	9,608.52	3,664.06	5,944.46	0.667
Chong qing	7,590.60	1,898.83	5,691.77	0.689
Gansu	6,945.02	1,435.58	5,509.44	0.630
Ningxia	5,698.61	289.96	5,408.65	0.674
Jiangxi	8,531.81	3,298.16	5,233.65	0.662
Guangxi	8,311.42	3,171.79	5,139.63	0.658
Hainan	1,588.12	271.62	1,316.50	0.680
Yunnan	13,085.16	14,416.76	−1,331.60	0.609
Qinghai	2,005.11	6,558.99	−4,553.88	0.638
Tibet	724.98	6,073.75	−5,348.77	0.569

Sources: The data on carbon sources and carbon sinks are derived from Niu Wenyuan, *China Carbon Balance Trading Framework Research* (Beijing: Science Press, 2008). HDI data have been extracted from the UNDP's *China Human Development Report 2013*.

The remaining 15 provinces and municipalities, barring Guizhou, belong to the fourth category. Their net carbon sources account for 26.18% of the global total net carbon source, and they are also regions characterized by a medium level of human development. Consequently, they fall within the group of proposed emission reduction regions. When they join the high HDI group (namely, the group of provinces with HDIs greater than 0.7 but less than 0.8), or even the extremely high HDI group (namely, provinces with HDIs greater than 0.8), they will shift

Table 5.4 Changes in regional levels of human development across China (1982–2010)

HDI group	1982	1990	2010
First World			Beijing Shanghai
Second World	Shanghai Beijing	Shanghai, Beijing, Tianjin, Liaoning, Guangdong, Zhejiang, Jiangsu, Heilongjiang, Jilin, Shanxi, Hainan, Shandong	Shandong, Jiangsu, Guangdong, Liaoning, Zhejiang, Heilongjiang, Inner Mongolia, Jilin, Fujian, Tianjin
Third World	Tianjin, Liaoning, Heilongjiang, Guangdong, Jilin, Shanxi, Hebei, Jiangsu, Zhejiang, Guangxi, Shandong, Hunan, Hubei, Inner Mongolia	Hebei, Fujian, Xinjiang, Guangxi, Hubei, Inner Mongolia, Hunan, Henan, Shaanxi, Sichuan, Ningxia, Jiangxi, Anhui, Gansu, Yunnan, Qinghai, Guizhou	Hebei, Shanxi, Henan, Hubei, Hunan, Anhui, Guizhou, Sichuan, Shaanxi, Xinjiang, Chongqing, Gansu, Ningxia, Jiangxi, Guangxi, Hainan, Yunnan, Qinghai, Tibet
Fourth World	Henan, Jiangxi, Fujian, Shaanxi, Xinjiang, Ningxia, Sichuan, Anhui, Gansu, Qinghai, Yunnan, Guizhou, Tibet	Tibet	

Note: Structures of human development are classified on the basis of the HDI values of various regions in 1982, 1990, and 2010. The Four Worlds classification in 2010 corresponds to extremely high, high, medium, and low levels of human development.

Data sources for the calculations were UNDP, *China Human Development Report* (1997, 2002, and 2012), China Financial and Economic Publishing House; National Bureau of Statistics, *China Statistical Yearbook 1983* (China Statistics Press, 1983); Population Census Office of the State Council & Department of Population Statistics, the National Bureau of Statistics, *1982 Population Census of China* (digital collection) (Beijing: China Statistics Press, 1985); and UNDP, *China Human Development Report 2013*.

from being proposed emission reduction regions to being conditional and unconditional emission reduction regions.

According to the data presented in the Stern report, the world's per capita emissions in 2005 amounted to 7 tons. In comparison, the per capita emissions of developed economies ranged from 10 tons in Japan to 22 tons in the United States. Per capita emissions in developing countries ranged from very small amounts emitted by the poorest countries to 2 tons emitted by India, up to 6 tons emitted by China. In view of this situation and the projected growth of the global population up to 9 billion, the global per capita emission amount should be 2 tons by 2050.

The latest World Bank data for 2014 reveals that China's per capita carbon emissions in 1990, 2000, 2006, and 2013 were 2.1 tons, 2.6 tons, 3.9 tons, and 7.2 tons, respectively. In comparison, the global per capita carbon emissions for these

Table 5.5 Per capita carbon emissions of China and
of the world (1990–2013)

Units: ton

Year	China	World
1990	2.1	4.1
2000	2.6	4.0
2006	3.9	4.3
2013	7.2	5.0

Source: World Bank database, 2014.

years amounted to 4.1 tons, 4.0 tons, 4.3 tons, and 5.0 tons, respectively. According to the emission reduction roadmap developed for China (Table 5.5), it is very likely that China's per capita emissions will peak in 2020. Assuming that China's population will reach 1.4 billion by 2050, it is hoped that the country's per capita carbon emissions in 2050 will be about 2 tons, which is the amount of emissions during the 1990s, or even lower.

The second constraint that compels China to undertake its emission reduction obligations arises from the fact that China is currently the largest emitter of CO_2, globally, and its per capita carbon emission has exceeded that of the entire EU region. Consequently, China should be obliged to assume responsibility for its emission reduction. The quantity of global carbon emissions resulting from human activities reached 36 billion tons in 2013. The largest proportion of this amount was contributed by China, accounting for 29%, followed by the United States (15%), Europe (10%), and India (7.1%). While China's per capita carbon emissions (7.2 tons) are still much lower than those of the United States and Australia, they have overtaken the quantity of per capita carbon emissions in the EU (6.8 tons) for the first time. Therefore, China must fulfill its required obligations.

5.3   The domestic context of China's emission reductions

5.3.1   *Changes in the mode of economic development (1996–2008)*

In 1995, it was proposed in the 5th Plenary Session of the 14th Central Committee of the Communist Party of China (CPC) to achieve the economic growth mode transformation. We succeeded in the beginning, and then the economic growth mode reversed.[2] During the period of the 9th Five Year Plan (1996–2000), China's economic growth rate was 8.63%. However, its energy consumption growth rate was only 1.10%, and the elastic coefficient of energy consumption demand was 0.127. During the period from 2001 to 2008, the development pattern was reversed. The economic growth rate was 10.2%, which increased by just 1.6%. However, the energy consumption growth rate was 9.4%, and the elastic coefficient was 0.922 (Table 5.6).

Table 5.6 China's energy, electricity, and coal consumption growth and elasticity coefficients (1996–2013)

	1996–2000	2001–2008	2009–2013
GDP growth rate (%)	8.630	10.200	8.400
Energy consumption growth rate (%)	1.100	9.400	5.160
Coefficient of elasticity	0.127	0.922	0.614
Generating capacity growth rate (%)	6.110	12.500	10.000
Coefficient of elasticity	0.708	1.225	1.190
Coal consumption growth rate (%)	−0.810	10.800	3.480
Coefficient of elasticity	−0.094	1.059	0.414

Source: *China Statistical Yearbook 2014*, World Bank.

A reversal of China's economic growth mode directly impacts on global energy consumption and pollution emissions. According to the UNDP *Human Development Report 2014*, compared to last year China's CO_2 emissions increased by 3.1% in 2012, accounting for 26% of the world's total emissions, and considerably exceeding the proportion of emissions contributed by the United States (12%). Fossil fuels accounted for 88.3% of China's primary energy supply, and the amount of per capita CO_2 emissions was 6.2 tons. During the period from 1970 to 2010, the average annual rate of per capita CO_2 emissions was 2.9%. In 2012, per capita CO_2 emissions had increased threefold compared with the amount in 1990. China has thus become the world's largest "black cat," responsible for the most negative externalities, globally. Consequently, China needs to change its development mode from black to green.

5.3.2   *China's prioritization of climate change mitigation*

The Chinese government attaches great importance to addressing climate change. As early as 1993, the National Climate Change Coordination Group was established. In 1994, the national *China Agenda 21* first proposed the concept of adaptation to climate change. Subsequently, in 2007, the 17th CPC National Congress report clearly proposed to "strengthen capacity building to address climate change, and make new contributions protecting the global climate." In a reversal of the economic development model articulated in the 10th Five Year Plan, the Chinese government explicitly proposed "energy conservation" targets and quantitative indicators. At that time, China had a very limited understanding of climate change, based on available knowledge and information sources, and failed to set a cap for CO_2 emissions as the core indicator. During the period of the 11th Five Year Plan, China did not achieve its targets relating to economic restructuring, such as increasing the proportion of added value of service industry in relation to GDP and the proportion of service industry in relation to total employment. This indicates that China's overly heavy industrialization had become the root cause of high energy consumption and high pollution during this period. **In a certain sense, this** could be viewed as a return to the heavy industrialization of the period from 1953 to 1959, with an even higher proportion of heavy industry in relation to total industry than the proportion that existed during that period.

5.3.3 Incorporation of climate change factors in major projects during the 12th Five Year Plan

In 2010, China's 12th Five Year Plan explicitly recommended "fully taking account of climate change factors in the planning, design, and fully taking account of climate change factors in productivity distribution, infrastructure, and major projects' designing." The plan further recommended "increasing the adaptation of agriculture, forestry, water resources and other key areas as well as coastal areas and ecologically fragile areas to climate change." In 2012, China issued a "National Plan for the Development of Science and Technology on Climate Change" during the 12th Five Year Plan Period to strengthen China's prowess in the science and technology fields in addressing climate change, narrowing the gap in relation to the advanced international level. The plan also aimed to promote China's emission reduction and adaptation to climate change, as well as to support technological innovation and application and the implementation of a sustainable development strategy in China.

In 2013, China's CO_2 emission per unit of GDP decreased by 28.5% from the unit amount in 2005, while the proportion of non-fossil energy in relation to primary energy sources increased to 9.8%. The country's capacities for hydropower and wind power, its scale of construction relating to nuclear power and solar collectors, and its number of rural biogas users ranked highest globally. Moreover, its area of forest land increased to 21.6% from 18.21% in 2005.

In May 2014, the 2014–2015 Action Plan on Energy Conservation, Emission Reduction and Low-Carbon Development, promulgated by the Chinese government, specifically proposed the target of decreasing CO_2 emissions per unit of GDP by more than 4% and 3.5% during the subsequent two years, respectively. The "National Plan on Climate Change for 2014–2020," introduced in September of the same year, proposed the following target to be accomplished before 2020: "Increasing the share of non-fossil fuels to around 15% of primary energy consumption, increasing forest cover by 40 million hectares and forest stock volume by 1.3 billion cubic meters compared with the 2005 level, lowering carbon dioxide emissions per unit of GDP by 40–45% compared with the 2005 level, and decreasing the carbon dioxide emission per unit of industrial added value by 50% compared with that in 2005."

5.3.4 Pressure from the international community

China needs to develop pro-peace and green development policies to respond to pressure being exerted by the international community while also actively participating in global actions to reduce poverty. For China to accomplish basic modernization by 2050, what is required is not just over 30 years of a peaceful international environment, but also a stable global climate and environment during this period. Cooperation with the United States and other countries, and leading and jointly promoting actions to reduce global emissions, constitute both opportunities and responsibilities for China.

Investment, even long-term investment, is necessary to address global climate change, and this investment ratio is much lower, relatively, than what is commonly imagined. The UNDP report notes that countries determine the agreed threshold of dangerous climate change as follows. At the end of this century, the average global temperature rise should be no more than 2 °C above the pre-industrialization temperature, and the atmospheric concentration of CO_2 should be stabilized at 450 ppm. This requires a decrease in global GHGs emissions by 40%–70% by 2050 compared with the amount of these emissions in 2010, and achievement of zero emissions by the year 2100. To achieve this goal, the balance of the global carbon budget should be around one trillion tons, of which half has already been used. If no definitive action is taken, the average global temperature rise by the end of this century is likely to exceed 4 °C. The results of such a rise would be catastrophic.

A globally acceptable emission pathway is required, the target of which is to reduce GHG emissions by 50% by 2050 compared with their levels in 1990. Developed countries need to fulfill their commitments during the current commitment period of the Kyoto Protocol, and to further agree to reduce GHG emissions by at least 80% by 2050, and by 20%–30% in 2020. In developing countries, the target of the main emission-producing countries is to reach their highest emissions by 2020, and subsequently to reduce this amount by 20% before the year 2050. The fifth IPCC report notes that countries only need to use 0.12% of the average GDP and can curb global warming from 2015 onward. The UNDP report estimates that to prevent carbon emissions from rising to dangerous levels, the world must spend about 1.6% of its GDP each year up to 2030. Developed countries currently have sufficient financial resources to invest in emission reduction technologies. They should, therefore, immediately implement emission reduction measures. At present, it may not be necessary for developing countries to undertake emission reduction obligations. From this index, it is evident that China has both the capability and the financial resources to make long-term investments to address climate change. In the future, avoiding excessive economic growth can prevent ups and downs in the economy and can also avoid excessive energy consumption and polluting emissions.

In setting forth the main objectives of China's socialist modernization process to be achieved by 2020, we assume that for the period of 2006 to 2020, China's average annual GDP growth rate will remain at 7.5%–8.0%, and that the GDP in 2020 will be equivalent to 4.65–4.98 times higher than that in 2000, with China achieving in advance or exceeding the goal of quadrupling its GDP. The energy consumption per unit of GDP in 2020 will further decline, and major pollutants will decrease by 10%–20%. Considering China's perspective, we propose midterm sustainable development targets that entail self-discipline or self-control, along with the application of resource and environment indicators,[3] including three top-priority indicators. By 2020, unit energy consumption will further decline, and major pollutants will decrease by 10%–20%. After 2030, the absolute amount of China's energy consumption is likely to show a downward trend, with energy consumption per unit output being reduced by half. The country's total water consumption will not increase, and a further decrease will occur in the proportion of water used for agriculture. The proportion of the society's overall investment in environmental protection in relation to GDP will increase from 1.3% to 2.5%, and emissions of major pollutants will be reduced by more than one-third. The forest cover rate will reach 23%–24%. This projection includes three secondary priority indicators: the proportion of renewable energy consumption, accumulated growth rate of forest stock, and the proportion of total pollution control investment in relation to GDP (Table 5.7). Consequently, China will advance toward the status

Table 5.7 Recommended primary and secondary nationally prioritized resource and environment indicators in 2020

Resources and environment indicators	*Unit*	*Attribute*	*2000*	*2005*	*2013*	*2020*
Primary priority indicators						
Energy consumption per unit of GDP[①]	Tons of standard coal/RMB 10 thousand	Constrained	1.45	1.54	0.76	0.74
Arable land inventory	100 million hectares	Constrained	1.32	1.22	1.22	1.18
Water consumption per unit of GDP[②]	Cubic meters/ RMB 10 thousand	Constrained	615	407	129	117
Forest cover	%	Constrained	16.55	18.20	21.63	23.40
Decline in major pollutants	%	Constrained				15–20
Secondary priority indicators						
Proportion of renewable energy consumption	%	Expected		5	13	15
Accumulated growth rate of forest stock	%	Constrained		>10	9.01	>30
Proportion of total pollution control investment to GDP	%	Constrained	1.00	1.30	1.68	>2.50

Notes: Data for 2000, 2005, and 2013 are from the *China Statistical Yearbook*, and data for 2020 are estimated by the author.

[①] Energy consumption per unit of GDP, and the GDP in 2000 and 2005 have been calculated in accordance with constant prices in 2000, and GDP in 2012 has been calculated in accordance with constant prices in 2010.

[②] Water consumption per unit of GDP has been calculated based on constant prices in 2000.

Source: Hu Angang: *2020 China: Building a Comprehensive Well-Off Society* (Beijing: Tsinghua University Press, 2007), 59, 62.

of being a resource-saving and environment-friendly society. We believe that by 2050, China will have the ability to accomplish the target set for developing countries of a 20% reduction in GHG emissions, and is even likely to achieve a sharp reduction.

One main area that China needs to invest in to address climate change is large-scale afforestation. Serving as a natural carbon sink, forests have huge potential for emission reduction. Citing the findings of a number of agencies, the Blair Report observed that 13 million hectares of forests (equivalent to the land area of Greece) were destroyed every year, and 24 million hectares of tropical forests were gradually being degraded. During the period from 2000 to 2005, Brazil and Indonesia alone accounted for half of the total damaged forest area. Varying degrees of forest degradation are occurring, but according to the Millennium Ecosystem Assessment scenarios, 200 to 490 million hectares of forests in developing countries will be destroyed by 2050, representing 5%–12% of the current total forest area across the globe.

Deforestation leads to a significant increase in GHGs, with the amount of CO_2 being 7.6 billion tons in 2000. This constituted 15%–20% of total GHG emissions. The Blair Report estimated that the potential emission reduction range of forests was 13 to 42 billion tons of CO_2 annually, while the price of carbon was US$100/ton of carbon dioxide equivalent (CDE) or lower. This potential can be realized in half of the forests, globally, at a cost that is less than US$20/ton of CDE. The world's reduction potential, estimated using a top-down model, is 9 to 14 billion tons of CO_2 annually.

A countercurrent to this forest degradation trend can be seen in China's emergence as the fastest-growing country, globally, in terms of forest resources. With the inception of the People's Republic of China, the country's forest cover rate has increased from 8.6% to the current level of 21.36%. The forest stock volume has reached 15.137 billion cubic meters, and the living forest stock volume is 16.433 billion cubic meters. Evidently, China has fundamentally reversed the "forest deficit" and become a "forest surplus" power.

Against the overall decline in global forest resources, China's forest resources continue to grow. The forest preservation area planted in China is nearly 67.7 million hectares, ranking highest globally and accounting for more than 50% of the total global area. China has achieved "double growth" of its forest area and stock volume. During the period from 2000 to 2005, the annual global reduction in forest areas was 7.3 million hectares, while forest areas in China showed an annual increase of 4.058 million hectares. The average annual global increase of planted forest areas was 2.8 million hectares, of which the annual increasing volume was 1.489 million hectares, accounting for 53.2% of the increase in planted forest area, globally.[4]

According to our preliminary calculations, because China's potential growth rate will decline in the coming decades, the annual cost for reducing emissions will have little effect on national economic development. According to the data presented in the Blair Report, and based on a scenario entailing China's development of a low-carbon economy, by 2040, China's GDP growth will be 6.9

times greater than it is now. In the absence of emission reduction, it will be 7.2 times greater than the current level. This estimate does not take into account the huge gains brought to China's economy by emission reduction and a low-carbon economy.[5] The estimate of China's emission reduction costs based on the time of implementation indicates that there is little difference in the impact on China's GDP, regardless of whether China begins emission reduction earlier (in 2010) or later (2030).

5.4 China's emission reduction roadmap and green modernization: A two-step strategy[6]

In the long run, public commitment to reduce emissions is concomitant with scientific development and the Chinese government's proposed construction of an ecological civilization. Moreover, the connotations, objectives, and processes entailed in this project are consistent with those established to ensure national energy security, address climate changes, and build a resource-conserving and environmentally friendly society.

At this stage, a very clear global emission reduction roadmap exists, entailing the following targets:

- By 2020, the CDE should reach peak value.
- By 2030, the annual emission should be less than 35 billion tons.
- By 2050, the annual emission should be less than 20 billion tons.

China has made a public commitment to reduce emissions and has announced its own emission reduction roadmap, involving major strategic decisions related to China's future long-term development. Essentially, it raises the following questions: whether China's national interests are consistent with those of human development, and whether the direction of China's development coincides with that of human development. China has made a clear commitment toward reducing emissions and currently faces the key challenge of how to implement and fulfill this commitment.

In 1987, the report of the 13th National Congress of the CPC proposed the following three-step concept:

> The first step is to double the Gross National Product (GNP) of 1980 and solve the problem of food and clothing for our people. This task has been largely fulfilled. The second step is to double the GNP again by the end of this century, thus enabling our people to lead a fairly well-off life (Xiao Kang level). The third step is to reach the per capita GNP level of moderately developed countries by the middle of the next century. This would mean that modernization has been basically accomplished and that our people have begun to enjoy a relatively affluent life. Then, on this basis, China will continue to advance.

Currently, the objectives of the three steps have basically been achieved, especially the second objective of a moderately prosperous society (Xiao Kang Society).

From a historical development perspective, we can posit that efforts to address climate change and to reach a global emission reduction agreement will be accompanied by the onset of a new industrial revolution – the green industrial revolution. China should seize this golden opportunity to become the leader, innovator, and driver of the fourth industrial revolution. Commencing from 1750, four industrial revolutions have occurred in the world. The first was the Industrial Revolution in Great Britain, which did not extend to China. The second was the American Industrial Revolution that occurred during the second half of the 19th century. Again, China missed the opportunity to be a part of this. The third industrial revolution was the Information Revolution that unfolded during the second half of the 20th century. This time, Chinese leaders were keenly aware of this great historical change and seized this important opportunity.

It is foreseeable that climate change will be one of the major international and domestic issues associated with China's ongoing development in the future. In this context, there are two pressing problems confronting Chinese leaders. The first concerns the question of how to transform China's current high-carbon economy into a low-carbon economy. The second issue relates to China's participation in global governance, entailing a shift from priorities of national governance to those of regional and global governance.

The theme and keywords of China's modernization process in the 21st century are "green development" and scientific development. This means changing from "black" to "green" in relation to all of the following: industrialization, urbanization, and modernization; manufacturing; energy; trade; cities; and consumption. Accordingly, China's green modernization process should follow a three-step strategy, described below.

The first step extends from the current time to 2020. The focus during this step is to mitigate CO_2 emissions and adapt to climate change. According to data released by the IEA, energy-related CO_2 emissions in China were 2.244 billion tons in 1990 and 5.684 billion tons in 2006. While China will strive to control its emissions at about 8 billion tons by 2020, it nevertheless accounts for about 20% of the total global CO_2 emissions. A significant reduction of the emission rate during the now-completed period of the 12th Five Year Plan (2011–2015) would be required, and the gradual stabilization of emissions during the period of the 13th Five Year Plan (2016–2020). By the end of the current plan, while agriculture will account for about 8% of the GDP, industry will drop to about 38%, and the service industry will increase to 47% of the GDP. The urban population will comprise 57% of the total population. Moreover, renewable energy sources will make up close to 20% of total energy sources, with coal consumption dropping to 60%, and the use of clean coal technology (especially CCS) being relatively high. The forest area will be 23%. Further, the HDI will reach 0.88, and national invention patent applications of domestic residents will rank third in the world. China's GDP is expected to rank second, while GDP (PPP) will rank first in the world.[7]

The second step, which is projected to occur from 2020 to 2030, will focus on reducing CO_2 emissions. By 2030, CO_2 emissions per unit of GDP will decrease to 60%–65% compared with emissions in 2005, first reaching a peak and then ceasing to increase. The proportion of low-carbon energy within the total energy consumption mix will increase to about 20%.[8] By that time, China's agriculture will account for just 5% of the GDP, with industry accounting for only about 30%. Conversely, the proportion of service industry will reach nearly 60%. The urban population will comprise 65% of the total population. In the energy sector, renewable energy will account for more than 25% of total energy sources, while the proportion of coal consumption will drop to 45%–50%, and clean energy usage efficiency will be very high. The amount of forest area will reach 24% of the total land area. HDI will reach a level of 0.93, while national invention patent applications of domestic residents will rank second in the world. Lastly, China will rank first, globally, for GDP and GDP (PPP).

The third step is projected to occur from 2030 to 2050. By 2050, CO_2 emissions will be reduced by 40%–70%, compared with the 2010 levels. GHG emissions should be reduced to zero in 2100 as a result of a substantial increase in the use of renewable and atomic energy as well as other low-carbon energy sources, promotion of energy-saving CO_2 capture and storage, and other initiatives. Alternatively, negative growth of atmospheric GHGs should be achieved through the application of capture and storage technology.[9] By this time, agriculture will account for only 2%–3% of China's GDP, the proportion of industry will decline to less than 20%, and the service industry will account for close to 80% of the GDP. The urban population will constitute more than 78% of the total population. In the energy sector, renewable energy will account for more than 55%, and coal consumption will drop to 25%–30%, thus achieving overall clean use. The forest area will have increased to 26% of the total land area. HDI will reach 0.98, and national invention patent applications of domestic residents will rank first in the world. In sum, China will have basically attained green modernization, being at the same level as that of developed countries and making genuinely green contributions to human society.[10]

China's road to modernization is considered to be an innovative road, differing from the traditional development model that has led to the combined growth of both the economy and of GHG emissions following the launch of the Industrial Revolution in Great Britain in 1750.[11] However, China will explore an economy-unrelated green development mode, aimed at achieving economic growth while reducing GHG emissions.

Green modernization is essential for China. While developing a green economy and industries, industrial restructuring, investing in green energy, and promoting green consumption will crucially address global climate change, these actions will not affect China's long-term economic growth rate. Rather, they will significantly improve the quality of economic growth and social welfare, thus achieving a multifold win-win situation that includes economic development and environmental protection, ecological security, and adaptation to climate change. At the same time, China, as the country with the largest population as well as the highest aggregate economic volume and the most patents, is bound to make significant global contributions in addressing climate change, implementing a "450-ppm stabilization" program and reducing GHGs by half through peaceful and green development as well as engagement in international and green cooperation.[12]

Notes

1 Carbon source refers to the process, activity, or mechanism that leads to the release of CO_2 into the atmosphere. A carbon sink denotes the process, activity, or mechanism for accomplishing the removal of CO_2 from the atmosphere.
2 The "Proposal of Communist Party Central Committee on Formulation of the Ninth Five-Year Plan for Economic and Social Development and the Long-Term Perspective in the Year 2010" was adopted during the 5th Plenary Session of the 14th Central Committee of the CPC on September 28, 1995. This stated: "To achieve the goals of the '9th Five Year Plan' by 2010, the key is the implementation of two fundamental changes with global significance. The first is a change in the economic system from a traditional planned economy to a socialist market economic system. The second is to change the economic growth mode from extensive to intensive, and promote sustained, rapid and healthy development and overall social progress."
3 Hu Angang, *2020 China: Building a Comprehensive Well-Off Society* (Beijing: Tsinghua University Press, 2007), 57, 59.
4 State Forestry Administration, *China Forestry and Ecological Construction Bulletin,* January 20, 2008.
5 Tony Blair, "Breaking the Climate Deadlock: A Global Deal for Our Low-Carbon Future," *Report submitted to the G8 Hokkaido Toyako Summit* (2008).
6 This refers to Deng Xiaoping's proposal of a strategic "three-step" conceptualization of China's modernization process. Accordingly, China was projected to reach the level of moderately developed countries by 2050. See Deng Xiaoping, "Proceeding from the Reality of Primary Stage of Socialism," in *Selected Works of Deng Xiaoping,* Volume III (Beijing: People's Publishing House, 1993), 251.
7 This also indicates that the implementation of emission reduction targets does not affect China's overall objective of building a moderately prosperous society by 2020. China's overall economy ranks first in the world, and its HDI places it in the global high-level group.
8 According to data presented by the IEA, China's CO_2 emissions in 1990 amounted to 2.2 billion tons and increased to 5.2 billion tons in 2005. (IEA, 2007).
9 IPCC Fifth Assessment Report (2014).
10 During a meeting held with Czechoslovakian Prime Minister Lubomir Strougal on April 26, 1987, Deng Xiaoping pointed out that by the middle of the next century, China could reach the level of moderate development. If this step is reached, it first of all would signify the accomplishment of a very complex and difficult task. Second, it would be a genuine contribution to human society. Third, it would more effectively reflect the superiority of the socialist system. Deng Xiaoping, *Chronicle of Deng Xiaoping (1975–1997)* (Part II), compiled by the Party Literature Research Center of the Chinese Communist Party Central (Beijing: Literature Publishing House, 2004), 1182.
11 The IEA report indicates that both the global GDP value in US$ and carbon emissions showed an increasing trend during the period from 1750 to 2007. See International Energy Agency, *The Impact of the Financial and Economic Crisis on Global Energy Investment,* 2009.
12 To ensure that the global temperature rise is no more than 2 °C by 2020, GHG emissions should be reduced by 25%–40% compared with levels in 1990. The objective of reducing overall emissions by 50% will be achieved in 2050. See page 263, Yang Jiemian, *Global Climate Change Diplomacy and China's Policy* (Beijing: Current Affairs Press, 2009).

6 Addressing climate change and achieving a low-carbon economy

The road ahead for China

A low-carbon economy (LCE),[1] associated with a new economic development model, is required to effectively address climate change. Developing an LCE as a global economic trend is the only way to deal with energy, environment, and climate security challenges. Such an economy is also appropriate in the context of China. It provides the country with an opportunity not just to extricate itself from an overreliance on carbon-based fuels, reduce the impact of oil price fluctuations on its macroeconomy, and achieve economic transformation, but also to maintain an appropriate and rapid rate of economic growth. China's implementation of an energy conservation policy began in the 1990s with the clear objectives of enhancing the nation's ability to adapt to climate change, and developing a national strategy for addressing this issue, articulated in the 12th Five Year Plan. The 18th National Congress of the CPC in 2012 further emphasized constructing an ecological civilization, adding new requirements for adaptation to climate change. China's carbon emission trading market is currently the second largest in the world, after that of the EU, and implementing carbon trading will help China to further achieve emission reduction targets and progress toward an LCE.

6.1 A low-carbon economy is essential for addressing the challenge of climate change

Energy security, environment, and climate change have become the top issues on the agendas of high-profile political meetings at the international level. The intertwining of these three issues has replaced security, in the traditional sense, giving rise to a new global issue. Dependence on and increased demand for fossil fuels have led to rising energy prices, particularly oil prices. High oil prices have caused civil unrest in oil-consuming countries, triggering intensified competition for oil resources and resulting in political turmoil in oil-rich regions such as the Middle East and Africa. Another consequence of excessive dependence on fossil fuels has been environmental destruction and adverse climatic impacts through excessive emissions. Scientific evidence presented in IPCC's Fifth Assessment Report on Climate Change has made it clear that human beings, through their activities, bear more than half of the responsibility for global warming. The credibility of this conclusion can be ascribed a value of more than 95%.

Developing an LCE is the only way to deal with security challenges relating to energy, environments, and the climate. The LCE model can enable us to effectively respond to climate change and achieve sustainable economic and social development. In 2003, the British government first proposed the concept of an LCE in its Energy White Paper, *Our Energy Future – Creating a Low-Carbon Economy*.[2] The overall goal described in this paper was to achieve a 60% reduction in CO_2 emissions by 2050 compared with the amount in 1990, fundamentally transforming Britain's economy into an LCE. The concept of an LCE was originally proposed in response to the concerns of developed countries relating to energy security and climate change. Following the release of the IPCC's Fourth Assessment Report on Climate Change and the *Stern Review on the Economics of Climate Change*,[3] the concept of an LCE has been widely accepted by nations across the world as the only effective way to tackle energy, environment, and climate change challenges and achieve sustainable development. For nearly the entire 21st century, for rich (developed), poor (developing), large, or small countries, LCE development will be a common approach for meeting the needs of economic transformation and sustainable development. As a result, this approach is directing the global economic trend.

Let us consider the two components of the LCE concept. Low-carbon entails minimizing economic development or curbing reliance on carbon-based fuels to achieve a transformation of energy and economies. This is a new development concept that contrasts with traditional patterns of high-energy consumption and high pollution. Economy means maintaining stable and sustainable growth on the basis of, and in the process of, transforming energy use. This concept does not exclude maximizing development and its outputs; nor does it reject the prospect of long-term economic growth. China is a major energy consumer and CO_2 emitter, and is also the largest developing country. Establishing an LCE would be well suited to China's national circumstances and development conditions. Considering the ideas "low-carbon" and "economy" encapsulated in this concept, China would accordingly need to stem its overreliance on carbon-based fuels, reduce the impact of oil prices, and achieve economic transformation. It would also need to maintain appropriate and rapid economic growth while solving a multitude of development issues.

Thus, the road to an LCE is not a smooth one. There are many people who are misinformed about the heavy reliance on fossil fuels, voicing skepticism about the reality of climate change. Many politicians are myopically focused only on "temptation" and "deterrence," driven by the ballot box, consequently failing to assume the broader responsibilities that their offices entail. They do not wish to carry out economic and social reforms because they do not want to lose the support of donors and special-interest groups who would be affected by such reforms. Fearing that their engagement with this topic will increase political pressure on them, they are reluctant to participate in deeper and broader social discussions. These politicians do not lack a keen strategic vision; rather, their decisions are largely attuned to the positions of interest groups. However, the raison d'être of politicians is to play a leading role on behalf of their societies at a time when their countries, and the world, are at a vital crossroads.

The former American vice president Al Gore raised an outcry over the attitude of the American government and society on energy and climate change. He stated:

> There are times in the history of our nation when our very way of life depends upon dispelling illusions and awakening to the challenge of a present danger. In such moments, we are called upon to move quickly and boldly to shake off complacency, throw aside old habits and rise, clear-eyed and alert, to the necessity of big changes. Those who, for whatever reason, refuse to do their part must either be persuaded to join the effort or asked to step aside. This is such a moment. The survival of the United States as we know it is at risk. And even more – if more should be required – the future of human civilization is at stake.[4]

Gore, who was a 2007 Nobel Peace Prize winner, vividly portrayed the embarrassing situation regarding the position of the United States on energy, environment, and climate security. He noted that the United States was borrowing money from China to buy oil from the Persian Gulf to burn it in ways that were destroying the planet. The dilemmas faced by the United States are in fact faced by most countries. They include high oil prices, excessive emissions, a fragile economy, and tense domestic relations. The program developed by Gore for the United States was to curb reliance on carbon-based fuels and achieve the goal of 100% renewable and truly clean electricity within 10 years. While this program may not be suitable for all countries, Gore's concept and his courage to state that "every bit of that has got to change" may offer lessons for countries worldwide, especially large developing countries such as China.

6.2   A low-carbon economy as the prevailing trend in global economic development

Addressing climate change and developing an LCE, while attributable to international pressure, offers China a unique opportunity to transform its mode of economic development, simultaneously providing the basic direction and driving force for this transformation. In the 21st century, human development is at the core of global development, and green development is the prevailing theme of human development. This emphasizes integrating and coordinating economic development and environmental protection, offering a more positive and people-centered road to sustainable development. LCE represents the future trend of global economic development. Participation in international negotiations, developing international rules in response to climate change, and developing an LCE are of tremendous international significance for China. China's engagement in these efforts will mark the first time that the country is taking such an initiative and actively participating in formulating international rules, thus directly influencing the new pattern of global development.

Let us recall the development of the modern world after World War II. During the second half of the 20th century and the early 21st century, the United Nations Charter and GATT (later to become the WTO) formulated the basic rules of the international political economic regime that all countries were required to follow. Similarly, to address the current challenge of global climate change, a new economy (LCE) needs to be developed and new international rules need to be drafted for GHG emission reduction, requiring the compliance of all countries. This implies that no country or region can undertake development that endangers the development of other countries or regions. Development should be conducted responsibly not just for our contemporary era, but also for history and the future.

Commencing in 1978, China's economy has entered an era of growth. However, this has been premised on China's passive acceptance of, and compliance with, the established international rules prescribed by Western countries and led by the United States. Today, 30 years after China's reform and opening up, the country's economic prowess and international status are unrecognizable from what they were in 1978. UN Secretary-General Ban Ki-moon has observed that China is now a global power, which entails global responsibilities. Consequently, China needs to actively participate in efforts to address a number of key global issues, including international peace and security, the Millennium Development Goals, climate change, human health, environmental protection, and sustainable development, and to cooperate with the UN. Thus, preserving resources and energy, protecting the environment, and reducing GHG emissions are indicators of China's self-restraint in relation to its economic rise, and are also in the fundamental interests of all of humanity.

China's decision to develop an LCE, and especially its decision to reduce its emissions, is not a simple technical decision. Rather, it is an important and strategic international political decision that relates to all of humanity, extending beyond a purely domestic strategic decision.[5] China's choices regarding climate change are vitally important. A mistake of a magnitude of merely 1% in China's decision is likely to cause a complete failure (100%) in the global response to climate change.[6]

The overproduction that is characteristic of East Asian emerging economies such as China and the characteristic overconsumption of the United States cause substantial economic imbalances. In this context, the rising prices of energy and other international commodities are causing concerns about inflation, forcing the United States to raise interest rates, while consecutive hikes in rates increase loan costs, lowering people's investment expectations and bursting the real estate bubble. This situation triggered the subprime mortgage crisis, and the chain reaction set off by this crisis affected global financial markets and real economies. The author believes that the roots of this financial crisis lie in overconsumption in the United States and overproduction by China. This imbalance is an extreme manifestation of unsustainable economic development, which indicates that the current global energy consumption and production model is not sustainable. There are two complex and closely linked chains underlying this situation.

> Chain 1: The East Asian type of production chain, which is centered on an energy use pattern that has fossil energy at its core
> Chain 2: The American type of consumption chain, which is supported by the US dollar–based global monetary system

The first chain determines that uncontrolled expansion of production, in the absence of coordination between producers, will inevitably result in high energy prices. The second chain determines that the expansion of production that is driven by consumption must show an uncontrolled spontaneous tendency under the "endorsement" of the hegemonic US dollar. In combination, these complex chains entail the inherent logic of self-deconstruction. The global political and economic structure, entailing a close linkage between these two complex chains, is constitutive of the "oil-dollar" hegemony.

It is apparent that the current issues of energy security and climate change are basically an outcome of the simultaneous tension and even fracture of these two chains. Focusing on one chain while neglecting the other may not be an effective solution. Thus, even with the full cooperation of China, India, and other emerging economies, the monetary and financial system will revert to its pre-crisis state and a new homogeneous crisis will again surface, engendered by the existing production and consumption chains in the foreseeable future. Therefore, it is essential for governance to address not just the dollar or oil issue, but also the "oil-dollar" that is linked to a hegemonic order. As long as energy derived from oil-based fossil fuels remains the principal axis of the world's energy production and consumption, the United States will not easily give up the oil-dollar linked hegemonic mechanism. Obtaining an adequate oil supply at a stable price is the most critical requirement for this mechanism.

The first oil crisis of the 1970s presented an opportunity to avoid an oil-centered production chain. Following this period, the international community has been inclined toward one of two approaches for stabilizing prices and obtaining a reliable energy supply. The first is the strategy pursued by the United States, which is to stabilize oil prices through collective actions and to coordinate supply and demand via negotiations. The second is the approach adopted by countries such as France to seek alternative energy sources and develop energy-saving technologies. The oil-dollar hegemony enables the former approach to become a seemingly cheaper and more reliable strategy. In particular, OPEC's (Organization of Petroleum Exporting Countries) integration of suppliers, and their compromises during this process relating to the oil-dollar hegemony, have provided the conditions for the future operation of the East Asian type of production chain based on an energy use pattern that has fossil energy at its core.

As long as established production and development patterns that are solely driven by fossil energy persist, the risk of escalating bulk commodity prices will continue to exist. China currently plays the role of a collaborator within the old order. On the one hand, this strengthens the control of the oil-dollar hegemony over supply and demand parties such as China. On the other hand, it means that China has to repeatedly play the role of "transfuser" or "payer" in case of the repeated occurrence of crises. For peripheral participants, such an order is an entrenched evil order imposed by core countries. If in the "Scramble" scenario, China and other countries that lack coordination can still seek temporary relief regardless of the consequences, then in the oil-dollar hegemony world, China will face unending exploitative institutional oppression. China must not only provide domestic products and services to other countries on a continuous basis in exchange for income, but it must also simultaneously write off its financial claims during periodic crises to enable the writing off of the debts of the core countries.

The existing low-carbon technology is inadequate for supporting China's transition to a low-carbon society in the near future, and the country will face real and potential difficulties on the road to an LCE. However, the challenges for China brought about by energy security and climate change are far greater. If China does not resolve to transform its economy into an LCE, it may face greater future costs. Developing an LCE entails a fundamental change in the production chain of China and other East Asian emerging countries. The new production chain can no longer be at the expense of the environment. More importantly, it can no longer be at the expense of uncontrolled aggravated energy consumption and should eradicate fundamental dependence on fossil fuels. As expansion of production inevitably occurs within a controllable energy consumption limit, the phenomenon of excessive production will not occur during the current great transformation, even in the event of large-scale expansion of production. An LCE will not bring about soaring prices resulting from extensive energy consumption. In other words, the low-carbon energy use mode breaks the inherent link between the expansion of marginal costs and production, which occur in synchrony.

Thus, transforming the East Asian type of production chain, which is based on a fossil fuel–centered energy use pattern, and reforming the monetary system, which is currently underpinned by the American type of consumption chain in which the US dollar is hegemonic, should be given equal importance in the long run. Applying both of the above strategies may provide the most effective solution for the current energy and financial crisis, as well as may be the only way forward toward the rational development of energy use and economic development for China.

6.3   Bleak prospects for the Kyoto Protocol and opportunities and challenges entailed in an increasingly decentralized global carbon market

Over a period of years, two key but differing modes of the international climate regime and policy have gradually evolved in the global carbon market. Initially a top-down model was developed, entailing a unified international carbon market formed under the Kyoto Protocol and other similar international treaties. Subsequently, a bottom-up model evolved, entailing domestic or regional carbon markets designed by countries according to their economic structures and political will in terms of distribution and accounting. This was followed by docking and coordination efforts relating to the different carbon markets at the international level.

As the world's first legally binding international environmental agreement, the Kyoto Protocol came into effect in 2005. The protocol stipulates reducing GHG emissions during the first implementation phase (2008–2012) by 5.2% based on their level in 1990. Applying the principle of common but differentiated responsibility, the protocol divides the contracting countries into Annex I Parties (developed countries and countries in transition) and non-Annex I Parties (developing countries). Of these, the Annex I Parties shall respectively assume certain obligations for reducing emissions during the first phase. For example, the 15 EU countries should reduce their emissions by 8% based on 1990 levels, the United States by 7%, and Japan and Canada each by 6%, or else face severe economic penalties. For the time being, non-Annex I parties are not required to undertake any emission reduction obligations. However, in 2001, the United States made it clear that it would not implement the Kyoto Protocol, and other developed countries followed suit. Canada exited from the Kyoto Protocol in 2011, while Japan, Russia, and other countries also made it clear that they would not participate during the second commitment period of the protocol. Similarly, countries such as Australia have refused to invest in the Green Climate Fund.

Therefore, the current focus of international climate negotiations has gradually shifted to developing a new global emission reduction agreement. At the same time, the size of the international carbon trading market, based on the mechanism under the Kyoto Protocol, is increasingly shrinking, while the global carbon market shows a decentralizing trend, and regional, subregional, and national-level carbon markets have been rising rapidly. Consequently, the parties have gradually adopted a bottom-up approach and have begun to seek docking by means of bilateral agreements with a view to achieving an integrated and unified carbon market in an international context.

An outstanding contribution of the Kyoto Protocol was introducing three flexible market mechanisms to address public environmental problems. Irrespective of their differing areas of focus on international emission trading, joint implementation (JI), or CDM, their common focus is on "overseas emission reduction." At its core, this entails quantifying CO_2 emission rights, enabling market trading and transactions, and guiding enterprises in obtaining the cheapest abatement costs worldwide. International emission trading refers to the mutual transfer of a part of the Assigned Amount Unit of CO_2 emission rights among developed countries, and JI means obtaining a cheap Emission Reduction Unit (ERU) through investment projects conducted among developed countries. Both of these require cooperation among developed countries. The Assigned Amount Unit and ERU not only can be used to offset emission reduction obligations, but can also be traded in international carbon markets. CDM refers to cooperation between developed and developing countries. Developed countries carry out interproject cooperation with developing countries through providing funds and technologies, and Certified Emission Reductions (CERs) achieved through such projects count toward the emission reduction commitments of contracted developed countries, as provided in the Kyoto Protocol. The CDM mechanism resolves the problem of high emission-reduction costs for developed countries and optimizes the allocation of resources on a global scale so that the costs of carbon emissions are greatly reduced. At the same time, it enables a good balance to be established between economic growth and low carbon-emission targets, and provides a step-by-step process for facilitating sustainable development in developing countries.

During the period following the Kyoto Protocol's entry into force in 2005, explosive growth of the global carbon trading market has been evident. According to the *Global New Energy Development Report 2014,*[7] the total quantity of carbon transacted in the global carbon market was 10.42 billion tons in 2013, and the total value of these transactions was approximately US$54.98 billion. In 2014, the world's carbon market reached a total value of US$32 billion. According to studies conducted by the World Bank and Ecofys, the total value of carbon pricing initiatives implemented in 2015 amounted to nearly US$50 billion. It has been projected that by the end of 2020, the demand for UN offset carbon credits will increase to 920 million tons in relation to the original base value of 1.69 billion tons. The CER supply will amount to approximately 1.3 billion tons, and the ERU supply will be approximately 900 million tons. During the period from 2014 to 2018, the net demand of the EU Emissions Trading System (EU ETS) regarding the European Union Allowance (EUA) in the secondary market will reach 1.3 billion tons. By this time, the total value of global carbon trading is expected to reach US$3.5 trillion, surpassing the oil market to become the world's largest market.

China's 12th Five Year Plan clearly articulates the goal of "gradually establishing a carbon emission trading market." In 2013, China selected five provinces and two cities as pilot areas for developing and implementing a carbon market mechanism. The design of the system was similar to that of the EU ETS, entailing carbon emission trading under total volume control, while also accepting offset carbon credits generated from domestic voluntary carbon reduction projects. Currently, China's carbon market is the second largest in the world after that of the EU. In 2013, the overall volume of China's carbon market trading was about 1.082 billion tons. Guangdong Province was China's largest carbon trading market, and the world's second-largest carbon trading market, with a total annual quota of about 388 million tons.

Based on whether or not the international carbon trading markets are under the governance of the Kyoto Protocol, they can be divided into Kyoto and non-Kyoto markets. The Kyoto market mainly comprises the EU ETS and the CDM and JI markets. The non-Kyoto market includes the voluntary Chicago Climate

Table 6.1 An overview of the carbon trading market (2005–2011)
Unit: million USD

Year	EU-ETS	GGAS	Quota-based CCX	RGGI	World	Project-based World	Total
2005	7,808	59	3	–	7,970	2,937	10,907
2006	24,436	225	38	–	24,699	6,536	31,235
2007	49,065	224	72	–	49,361	13,646	63,007
2008	100,526	183	309	198	101,492	33,574	135,066
2009	118,474	117	50	2,179	122,823	20,913	143,736
2010	117,800	–	–	–	–	–	141,900
2011	148,000	–	–	–	–	–	176,000

Source: World Bank.

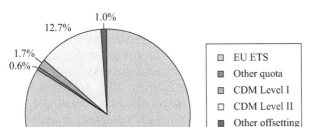

Total amount of carbon trading: US$176 billion

Figure 6.1 The global carbon market structure (2011)
Source: World Bank, *2013 China Carbon Financial Outlook.*

Exchange (CCX), the mandatory New South Wales Greenhouse Gas Abatement Scheme, and various retail markets.

The EU ETS was officially launched on January 1, 2005, as the EU's Cap-and-Trade Program for reducing GHG emissions. This program is the world's first multinational carbon emission trading system, and is also the carbon trading system with the greatest influence, globally. Member states submit their national allocation plans on CO_2 emissions for each phase, which are formally approved by the European Commission. They subsequently propose specific emission reduction targets for relevant enterprises and determine how to allocate emission rights to these enterprises. Because of the uneven economic development of countries, abatement costs differ, and the quotas may be sufficient or insufficient. Consequently, flexibility of EUAs becomes feasible as well as necessary.

Two phases of the EU ETS implementation plan have so far been completed. The first phase (2005–2007) mainly focused on the power sector and other energy-intensive industries, which accounted for 44% of the total amount of European emissions. During this phase, the European Commission annually issued 2.298 billion EUAs for the 27 member states (1 EUA = 1 tCO_2).[8] The second phase (2008–2012) coincided with the initial operational phase of the Kyoto Protocol. During this phase, the quota was lower than it was during the first phase, with an upper annual limit of 2.098 billion EUAs for the emissions of the EU-27 member states. Moreover, the national allocation plans decreased by 10.4% compared with the quantities initially declared by the member states.[9] In 2011, EU ETS carbon trading prices plummeted, but subsequently showed a slight rebound, stimulated by the EU's rescue measures and discussion of a reform plan. However, the overall downtrend persisted. Currently, the EU ETS is in its third phase of operation (2013–2020). Gradual implementation of a series of major reforms has begun, including "back-loading" and an adjustment plan for emission limits after 2020. Specifically, to respond to the negative factors of the current continuous downturn in the carbon trading market, and to low EUA prices over the long term, a back-loading strategy that is based on the premise of there being no reduction in the total quota supply proposes delaying the 900 million quota auction for the period of 2013 to 2015 to the period of 2019 to 2020 prior to a market launch. In addition, the EU ETS hopes to further reduce GHG emissions by 43% in 2030 from 2005 levels.

During the period from 2005 to 2010, the EU's huge demand promoted the rapid growth of the CDM market. However, with the overall decline of the Kyoto emission reduction mechanism, and the EU's progressive tightening of the CER application process, the CDM market gradually began to shrink. Following the establishment of the CDM market, China, India, Brazil, and South Korea have emerged as its major suppliers, collectively accounting for more than 80% of shares in the global CDM market. The buyers' market for CERs has always been dominated by developed countries (Europe and the United States). The EU carbon emission quota in 2011 accounted for 84.03% of the trading value of the global carbon market, which would otherwise have created opportunities for CDM projects within developing countries. However, to control the total CER trading volume, developed countries have limited substitutability between the ERU and CER, and have further controlled the pricing of CER. Nevertheless, according to UNDP data, developed countries achieved the targets of reducing GHGs emission amounting to 5 billion tons in 2012, half of which was realized in the form of CDM. China accounted for 35%–45% of the total global CDM potential in 2010, which was equal to the total potential of Latin America, Africa, and the Middle East. Thus, China's reduction of carbon emissions has been the greatest under the CDM.

Commencing from June 18, 2013, pilot projects for establishing carbon emission trading were successively introduced in Shenzhen, Shanghai, Beijing, Guangdong, Tianjin, Chongqing, and Hubei. Consequently, China's carbon trading entered a new era entailing a shift from reliance on projects accessed within the worldwide CDM market to the development of a domestic quota market. According to statistics provided on the UNFCCC website, as of August 31, 2014, China had successfully registered 3,805 CDM projects, issued primary Certificated Emission Reductions (pCER) equivalent to 627 million tons of CO_2 with buyers, and was ranked highest for the number of registered CDM projects and the volume of pCER issued. Currently, the participation of China's domestic enterprises in international carbon trading is scattered in the absence of a unified national carbon trading platform, mainly through the operation of CDM projects with apparent regional features. These projects are largely concentrated within low-end, low-tech, and energy-saving fields such as wind energy, hydropower, and biogas. Thus, China is not currently engaged in unilateral projects entailing a considerable degree of complexity and a requirement for advanced technology in emission reduction. In addition, China's CDM project enterprises are at the bottom rung of the international CDM industry chain and lack access to information on cost accounting as well as international carbon trading prices and trends. They also lack an awareness of the risk of costly repurchasing that they will face in the future.

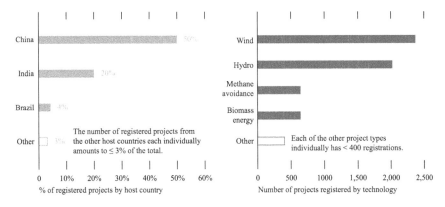

Figure 6.2 Number of registered CDM projects, worldwide, according to country and technology

Source: World Bank's *Carbon Market Situation and Trend Report 2014.*

By 2016, China plans to establish a national carbon trading market and will gradually reduce the proportion of free quotas. Moreover, it is expected to expand the defined scope of emission-controlling entities and include more market entities. This will promote the development of the carbon emission market, as well as energy conservation by enterprises in a more market-oriented manner. Although China is the largest primary CER supplier, because capital projects are not freely convertible in RMB, the issue of the settlement currency poses difficulties. Given that they do not bear responsibility for emission reductions, the only avenues for participation in international carbon trading for China and other developing countries are the CCX and other voluntary markets.

Compared with the CDM mechanism that underwent a rise, followed by a decline, the JI mechanism has not been effectively promoted for the long term, and uncertainty regarding the future has further limited the ERU supply. According to World Bank statistics, in 2012 – that is, before the end of the first commitment period of the Kyoto Protocol – the total global amount of ERU was 526 million tCO_2, of which Russia and Ukraine together contributed 492 million. In 2013, only 26 new JI projects were included within the framework of the UNFCCC, as compared with 229 projects added in 2012, indicating a decrease of nearly 90%. The increased projects in 2013 declined by 98% compared with the projects introduced in 2012. Moreover, because of the lack of supervision by the Joint Implementation Supervisory Committee, many projects cannot actually be implemented. Consequently, the future prospects of this mechanism are highly uncertain.

As previously noted, non-Kyoto markets include the CCX, the New South Wales Greenhouse Gas Abatement Scheme, and retail markets. Currently, America's carbon emission trading is mainly implemented voluntarily by means of the "total and control" method based on the CCX, and implemented mandatorily based on the Regional Greenhouse Gas Initiative (RGGI), applying this same method. In 2003, the CCX, which is the second-largest carbon trading market in the world, and the only market conducting emission trading in six GHGs (CO_2, CH_4, N_2O, hydrofluorocarbons [HFCs], perfluorinated compounds[PFCs], and sulfur hexafluoride [SF_6]), established the world's first voluntary emission-trading market. The RGGI is the first mandatory CO_2 emission trading market project to be implemented in the United States. It is a cooperative program developed by 10 northeastern and mid-Atlantic states to limit GHG emissions. Regulated power plants may use the CO_2 quotas of any of the member states to comply with the RGGI agreement. During the early period of its operation, RGGI faced a serious problem of oversupply. Carbon prices were in the doldrums over the long term, and trades were highly infrequent. In 2013, RGGI proposed a reform program to introduce total quota austerity and change the cost control mechanism. Beginning in 2014, the annual quota was slashed by more than 45%. This program significantly reactivated RGGI's carbon market, resulting in a steady rise in market prices.

The major carbon exchanges of the EU ETS include the European Climate Exchange, Powernext, Nord Pool, the European Energy Exchange, the Energy Exchange Austria, and Bluenext. Of these exchanges, the European Energy Exchange and Energy Exchange Austria mainly deal with EUA spot trading, publishing these trading prices on a daily basis. The European Climate Exchange leads in trading on new carbon financial instruments, offering EUA futures and options. During the period from 2005 to 2012, the main products traded were EUA contracts delivered in December each year. By contrast, the trading scales of the dollar-denominated CCX, the Chicago Climate Futures Exchange, and the New York Mercantile Exchange, which have all launched carbon futures and options, are small. There is no doubt that the euro is the main pricing and settlement currency relating to carbon stock and carbon derivatives trading. Off-exchange trading is also very common. The UK has always been the most staunchly committed carbon-emission reducer. Although the UK Emission Trading Scheme has been incorporated into the EU ETS, London's position as a global carbon-trading center has been firmly established, and there may still be room for the British pound to be used in carbon trade pricing and as the settlement currency.

Although the Kyoto Protocol is named after a Japanese city, and Japan also plays an important role in promoting and implementing carbon trading, this country has demonstrated a slow pace of action compared with that of Europe. In 1997, Nippon Keidanren launched Keidanren Voluntary Action Plans, covering 82% of the emissions of 34 Japanese industries. These plans constitute a market system that is mainly determined by the Kyoto Target Achievement Plan to which they are linked. In October 2008, Japan began to implement its voluntary emissions-trading scheme. Based on the *Outline for Promotion of Efforts to Prevent Global Warming,* released in 1998, The Kyoto Protocol Target Achievement Plan was successively modified in 2002, 2005, and 2008. Each of the modified plans described the introduction of a new emission-trading plan for Japanese industry that would integrate the Keidanren Voluntary Action Plans with other existing initiatives, as well as develop a pilot voluntary emission-trading system. At the G8 Summit held in 2008, the Japanese Prime Minister, Yasuo Fukuda, announced Japan's long-term GHG emission targets. Accordingly, Japan would reduce its GHG emissions by 14% by 2020, and by 60%–80% by 2050 compared with 2005 levels. Notably, the emission reduction target for 2020 is the same as that proposed by the EU for the same year. At the level of exchanges, an experimental Japanese emission-trading system, designed by the Japan Electric Power Exchange, was officially launched in October 2008. In April 2010, the Tokyo metropolitan cap-and-trade system was officially launched as Asia's first carbon-trading system. This was not only Japan's first region-wide cap-and-trade system, but it was also the world's first municipal cap-and-trade system. The system covers more than 1,400 sites (including 1,100 commercial facilities and 300 plants), accounting for 20% of Tokyo's total emissions. The Tokyo Metropolitan Government has established a GHG reduction target of 25% for 2020 compared with the amount in 2000.

In 2011, Japan announced its withdrawal from the second commitment period of the Kyoto Protocol and relaxed the restrictions on GHG emissions. This resulted in an increase in the country's GHG emissions by about 11% in 2013 compared with emission levels in 1990. In July 2015, the Global Warming Prevention Headquarters, established under the Japanese cabinet, formally finalized "Japan's Commitment" (Draft), delineating GHG emission targets for the period of 2020 to 2030. Accordingly, the new target set for decreasing Japan's GHG emissions by 2030 was 26% of the level in 2013. However, the fact that Japan announced its withdrawal from the second commitment period of the Kyoto Protocol in 2011, and its failure to strictly enforce the GHG reduction program established for the period from 2013 to, must be considered. Consequently, it is estimated that by 2030, the total amount of Japan's new GHG emissions will actually decrease by only 18% compared with the amount in 1990, which was established as the base year of emission reduction in the Kyoto Protocol.

Australia originally planned to implement a carbon price mechanism over two phases. The first phase (July 1, 2012 to June 30, 2015) was to be the fixed carbon price phase, and the second phase (after July 1, 2015) was the transitional phase toward total GHG control and establishing an emission-trading mechanism. However, after the Australian Conservative Coalition came to power in 2013, a proposal to abolish the fixed carbon price mechanism and the carbon tax were accepted by the House of Representatives and the Senate. On July 17, 2014, the Australian government announced the abolition of the carbon price mechanism, which had been in operation for only two years, putting an end to the possibility of its transition to a floating price phase. As the highlight of its direct climate action plan, the Australian government established an Emission Reduction Fund for GHG emissions comprising A$2.55 billion. The beneficiaries of these funds were enterprises that were willing to reduce their emissions through the purchase of emission reduction targets, as a replacement of the previous carbon price system. However, if the Australian government does not adopt the suggestion of a 36% emission reduction, proposed as an international climate change response, the Emission Reduction Fund can only help Australia to reach 11% of the 2000 emission reduction level by 2025, which is well below the post-2020 emission reduction targets to be established and announced by the Australian government. Moreover, according to a recent report issued by RepuTex, an Australian and research and rating agency, the Emission Reduction Fund will be used up within the next year. Accordingly, Australia's policy formulation for future responses to climate change is highly uncertain.

In 2013, the development of a carbon market at national and regional levels was evidently progressing steadily, with the establishment of eight new global carbon-trading markets (including the California Cap-and-Trade Program, the Quebec Cap-and-Trade Program, Kazakhstan's carbon emission trading market, and China's five carbon emission trading pilots in Beijing, Shenzhen, Shanghai, Guangdong, and Tianjin). In April 2014, China officially launched a carbon-trading pilot project in Hubei. This trading pilot, entailing a well-designed overall system and an improved compliance risk control system, quickly became the second-largest carbon market globally and the largest in China, accounting for 48% of the country's carbon market. In addition, Brazil, Chile, Thailand, Indonesia, and South Africa have announced their follow-up carbon emission trading policies, and Mexico will introduce a carbon tax system and emissions trading scheme, which will be a valuable complement to the global carbon market. Tunisia and Russia have also made progress in this area, with the signing of a presidential decree in Russia, which intends to establish a carbon-trading system in the future.

6.4   An energy-saving policy as an important way for China to move toward establishing a low-carbon economy

China has implemented the most proactive domestic energy-saving policy ever, and its domestic emission-reduction policies are synchronous with those at the global level. The Chinese government attaches great importance to climate change. Even around the time of the United Nations Environment and Development Conference in 1992, the Chinese government took the lead in formulating *China's Agenda 21 – White Paper on China's Population, Environment and Development in the 21st Century*. In 1998, China became a signatory of the UNFCCC, and it subsequently ratified the Kyoto Protocol in 2002. In recent years, the Chinese government has taken a series of measures to address climate change at the four levels of strategy (national policy), institution (legal and planning), organization, and mechanism. At the G-8 Summit's Leaders' Meeting of Major Economies on Energy Security and Climate Change, held in July 2008, President Hu Jintao emphasized the building of an ecological civilization as a strategic task. He also stressed "the need to adhere to the basic national policies of resource conservation and environmental protection, and strive to create an industrial structure, growth pattern, and consumption mode that saves energy and resources and protects the ecological environment." The 12th Five Year Plan has clearly stated the need to enhance the ability of China to adapt to climate change, and to develop a national strategy for this to occur. The 18th National Congress of the CPC also emphasized constructing an ecological civilization and added new requirements for adaptation to climate change. During the Asia-Pacific Economic Cooperation (APEC) meeting held in November 2014, China and the United States jointly issued the "U.S.-China Joint Announcement on Climate Change." In this document, China declared its intention to achieve the peaking of CO_2 emissions by around 2030 and to increase the share of non-fossil fuels in the country's primary energy consumption to around 20% by 2030. Through a combination of economic and social development planning and applying a strategy to promote sustainable development, China succeeded in establishing the National Leading Group on Addressing Climate Change, formulating and promulgating a series of laws and regulations, and taking a series of measures to address climate change. China has also actively promoted energy conservation as an entry point for addressing climate change, and has introduced a series of measures such as energy saving, optimizing the energy structure, improving energy efficiency, and promoting afforestation. These are remarkable achievements.

6.4.1 Strategic (national policy) level

The first measure at the policy level has been the recognition that scientific development is an important guiding principle for China's economic and social development. Development, the core of which is putting people first, is the first prerequisite for applying the concept of scientific development. This approach basically requires that development is comprehensive, coordinated, and sustainable, and its fundamental method is to take all factors into consideration.

The second measure has been the decision to construct an ecological civilization, viewed as an important strategic task. The *Report to the 17th National Congress of the Communist Party of China* deems the construction of an ecological civilization to be a long-term effort to enhance the welfare of the people and to secure the nation's future. It emphasizes that in the face of grim situations entailing tight resource constraints, environmental pollution, and ecosystem degradation, China must establish the concept of an ecological civilization based on respect for and protection of nature, and promoting harmony with nature. This requires that the construction of an ecological civilization is integrated into the construction of all aspects of the economy and of politics, as well as cultural and social construction.

The third measure entails the continued emphasis on resource conservation and environmental protection within basic national policies. Family planning and economic opening up have been established as two basic national policies. Following the release of the *Report to the 17th National Congress of the Communist Party of China,* it is evident that the Chinese government has included resource conservation and environmental protection as basic national policy directives, indicating the high prioritization of resource conservation and environmental protection by the CCP and the government of China. The *Report to the 18th National Congress of the Communist Party of China* (2012) further emphasizes the need to adhere to priorities of preservation, conservation, and a natural recovery-oriented approach, and to make efforts to promote green, cyclic, and low-carbon modes of development. The latter report also emphasizes developing spatial patterns, industrial structures, production modes, and lifestyles that promote resource conservation and environmental protection, reverse the trend of ecological deterioration at its source, create a good productive and living environment for people, and contribute to global ecological security.

6.4.2   Institutional level

As a first measure, in October 2008, the government of China announced the release of a white paper titled *China's Policies and Actions for Addressing Climate Change,* which clarified six principles: addressing climate change within the framework of sustainable development; adhering to the principle of "common but differentiated responsibilities"; paying equal attention to both emission reduction and adaptation; developing conventions and protocols as the main channel for dealing with climate change; relying on technological innovation and technology transfer; and promoting public participation, as well as extensive international cooperation.

As a second measure, in June 2009, participants at the meeting of the National Leading Group to Address Climate Change and Energy Conservation and Pollutant Discharge Reduction formally proposed to deem CO_2 emission reduction as an important development indicator. This laid the foundation for including CO_2 emission reduction as a core indicator in the 12th Five Year Plan.

A third measure entails China's promulgation of the National Climate Change Program in 2009, and its adoption of the "Resolution of the Standing Committee on Positively Addressing Climate Change" in August of the same year. In November 2009, the following target was set at the executive meeting of the State Council. By 2020, China's CO_2 emissions per unit of GDP will decrease by 40%–45% compared with the amount in 2005, and the proportion of non-fossil fuels in relation to primary energy consumption will reach about 15%. The forest area will increase by 40 million hectares in relation to the forest area in 2005, and forest stock volume will increase by 1.3 billion cubic meters compared with this volume in 2005.

The fourth measure is evident in the clear commitment articulated in the 12th Five Year Plan to enhance the country's ability to adapt to climate change and to develop national strategies to enable adaptation to climate change to occur. The 12th Five Year Plan explicitly noted that climate change factors should be fully considered in planning, design, and constructing the distribution of productive forces, infrastructure, and major projects. It further noted that efforts should be made to increase adaptive capacities of agriculture, forestry, water resources, and other key areas, as well as coastal areas and ecologically fragile areas, to climate change.

As a fifth measure, in 2011, the Chinese government introduced the *Work Plan for Greenhouse Gas Emission Control during the 12th Five Year Plan Period,* which provided a clear reduction target of 17% for the country's CO_2 emissions per unit of GDP by 2015 compared with the amount in 2010. It recommended efforts to control non-energy-related activities and emissions of CO_2 as well as those of methane, nitrous oxide, hydrofluorocarbons, perfluorocarbons, sulfur hexafluoride, and other GHGs.

Lastly, in November 2013, the Chinese government introduced the National Strategy for Climate Change Adaptation. This was based on a thorough assessment of the current and future impacts of climate change on China. It clarified general guidelines and principles for the country's adaptation to climate change up to 2020, and proposed adaptation objectives, key tasks, regional patterns, and safeguarding measures, thereby providing guidance for the overall coordination of adaptation measures.

6.4.3   Organizational level

The first measure taken at the organizational level has been creating the National Energy Leading Group to coordinate the ministries and other government agencies. One of the most important responsibilities of this group, which has established a working office, is to develop energy strategies and coordinate the drafting of energy laws.

A second organizational measure has been the establishment of the National Leading Group to Address Climate Change and Energy Conservation and Pollutant Discharge Reduction. This group provides overall leadership in relation to responses to climate change and energy conservation.

6.4.4   A mechanism

The adopted mechanism entails making full use of emission trading to implement CDM projects, and increasing emission reduction credits. In 2004, the Chinese government initially approved four CDM project applications. As of May 5, 2015, the National Development and Reform Commission had approved a total of 5,073 CDM projects.[10] According to the statistics presented on the website of the CDM Executive Board, as of June 30, 2015, China had registered 3,762 projects, accounting for 49.2% of all projects registered worldwide (see Figure 6.4). It is expected that China will receive 595,971,672 tons of issued carbon emissions, representing 60.1% of the total issued amount (991,526,024 tons) of the CDM projects of host countries.

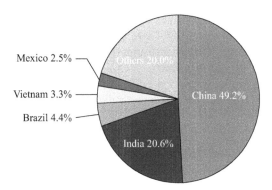

Figure 6.3 Proportion of China's registered CDM projects to the total number of projects registered worldwide (as of June 2015)

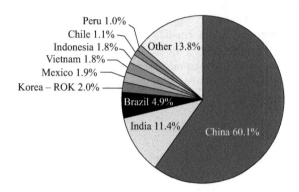

Figure 6.4 Country-wise distribution of expected annual Certified Emission Reductions received from registered projects according to the latest forecast (2015)

Source: http://cdm.unfccc.int/Statistics/Public/CDMinsights/index.html.

6.5 Reducing energy consumption per unit of GDP as the main emission-reduction method for China

The Kaya formula indicates that carbon emissions are dependent on the population, per capita GDP, energy consumption per unit of GDP, and emission per unit of energy consumption as shown below:

Emission = population × per capita GDP × energy consumption per unit of GDP × emission per unit of energy consumption

Jiasi and Chaoling (2008) have pointed out that China's birth rate has been below the global average, indicating a slow growth trend. China has a more pressing mission of eradicating poverty, which relates to its per capita GDP. This should not be reduced unless there is no other way to reduce emissions. Emissions per unit of energy consumption are usually considered as the emission factor, which is generally a stable value by default, unless a significant breakthrough in combustion technology occurs. Therefore, in the Chinese context, the only way to reduce emissions is to reduce energy intensity. There is considerable potential for improving China's energy efficiency through measures that mainly entail energy restructuring and energy efficiency improvement.[11]

A country's energy consumption per unit of GDP is calculated as the total amount of energy consumed during a particular year divided by the total amount of that year's GDP:

$$\text{Energy consumption per unit of GDP} = \frac{\text{Total energy consumption}}{\text{GDP}}$$

Many factors affect energy consumption, including land area and available natural resources; the climate and environment; the industrial and consumption structures; purchasing power; the technological structure; the energy supply structure; living habits, facilities, and organizational modes; energy prices; energy and climate policies; and management systems. Some of these cannot be changed, whereas others can be modified through human intervention. Therefore, energy-saving initiatives should focus on fields that are amenable to human intervention through economic, administrative, legal, and other means. Traditionally, two main approaches have been used to reduce energy consumption. The first is to reduce the proportion of high-energy consumption and low-value-adding industries in the national economy through industrial restructuring – that is, structural energy saving. The second is to reduce the energy density of various industrial sectors through technical means – that is, technological energy saving. Viewing both of these approaches as being very limited, Xinhua (2006) has proposed a new consumption reduction formula as follows:

$$\frac{E}{GDP} = \sum_i \frac{A_i}{GDP} \cdot \sum_j \frac{A_{ij}}{A_i} \cdot \frac{E_{ij}}{A_{ij}}$$

In this formula, A_i is a sectoral activity, such as household consumption, passenger transport, cargo transport, and manufacturing output; A_{ij} is a subsector's activity, such as car passenger transport, bus passenger transport, and air passenger transport within the passenger transport sector; and E_{ij} refers to the energy consumed during activities conducted within a subsector. On the right-hand side of the formula, $\frac{Ai}{GDP}$ refers to a particular economic sector's activity density in relation to GDP (such as the density of passenger transport per unit of GDP). $\frac{Aij}{Ai}$ expresses the structure of the sectoral activity (such as the ratio of private car passenger transport to bus passenger transport), and $\frac{Eij}{Aij}$ is the unit energy consumption of activities within subsectors of the sector (such as energy consumption per unit of private car passenger transport). This formula can be expressed as:

GDP energy density = sector activity density per unit of GDP × sector activity structure × energy density of subsector activity

Reducing energy consumption per unit of GDP requires a focus on implementing three measures. The first is reducing the intensity of an activity per unit of GDP. The second entails industrial structure adjustment of all sectors, including the consumer, transport, and manufacturing sectors. The third relates to adopting energy-saving technologies to reduce energy consumption per unit of activity.[12] In the process of promoting energy conservation, China also needs to carry out extensive and fundamental groundwork to clearly understand the theoretical basis of energy conservation and take more targeted policy measures. The emission reduction targets to be attained by 2015, as proposed in China's 12th Five Year Plan, are a decrease of 11.4% in the proportion of China's non-fossil fuels in relation to primary energy consumption, a decrease of 16% in the energy consumption per unit of GDP compared with the amount in 2010, and a decrease in CO_2 emissions per unit of GDP by 17% compared with the amount in 2010. Specifically, through dual control measures relating to energy consumption intensity and total energy consumption, China achieved a total energy consumption of 4 billion tons of standard coal by 2015, with energy consumption per unit of GDP decreasing from 81 tons of standard coal/million RMB GDP in 2010 to 68 tons of standard coal/million RMB GDP, demonstrating a cumulative decline of 16% within five years. Standard coal consumption for the supply of thermal power decreased from 333 g/kWh in 2010 to 323 g/kWh in 2015, and the average annual rate of decrease was 0.6%. The total rate of loss in relation to the comprehensive grid line was 0.2% between 2010 (6.5%) and 2015 (6.3%). Thermal power supply standard coal consumption decreased to 323 g/kWh, and moreover, energy consumption relating to comprehensive refiner processing dropped to 63 kilograms of standard oil/ton, and comprehensive energy efficiency increased to 38% between 2010 and 2015.

China's emission reduction targets are very challenging and difficult to achieve. The 12th Five Year Plan projected that CO_2 emissions per unit of GDP would decrease by 17% by 2015 compared with their levels in 2010. Further targets included a fall in SO_2 emissions to 1.5 grams per kWh of coal power, and of nitrous oxide emissions to 1.5 grams. Moreover, the emission intensity of fine particulate matter ($PM_{2.5}$) resulting from the development and use of energy was projected to decrease by more than 30%, and the reclamation rate of land under coal mines was projected to be over 60%. However, China's per capita primary energy consumption in 2011 was 2.6 tons of standard coal, showing an increase of 31% compared with the amount in 2006, while the per capita consumption of natural gas was 89.6 cubic meters (an increase of 110%), and the per capita electricity consumption was 3,493 kWh (an increase of 60%). In addition, the Chinese government has made a commitment that by 2020, non-fossil fuels will comprise about 15% of the primary energy consumed, and CO_2 emissions per unit of GDP will fall by 40%–45% compared with the amount in 2005. In general, there is a long time span from policy formulation to reflection on the actual outcomes of the policy. This also applies to the time span between introducing an energy-saving policy and reflections on its outcomes, which is relatively long because of the existence of inert infrastructure. By contrast, the EU has only committed to reducing energy consumption per unit of GDP by 20% during a 15-year period up to 2020.[13]

During the period of the 11th Five Year Plan, China made significant progress in adjusting its industrial structure, conserving energy, improving energy efficiency, and developing and using new energy sources. According to a white paper titled *China's Energy Policy (2012),* China's energy consumption per RMB 1000 of GDP dropped by 20.7% during the period from 2006 to 2011, with a savings of 710 million tons of standard coal. In 2011, the country's coal consumption for producing thermal power was reduced by 37 grams of standard coal/kWh (10%) compared with the amount of coal consumed in 2006. Moreover, China is actively developing new forms of energy and renewable energy. In 2011, the installed capacity of the country's hydropower reached 230 million kilowatts, ranking highest in the world. In the field of nuclear power, China also ranks highest globally, with the operationalization of 15 generating units with an installed capacity of 12.54 million kilowatts. A further 26 units are under construction, with an installed capacity of 29.24 million kilowatts. Moreover, China's on-grid wind power capacity also ranks highest globally, at 47 million kilowatts. Additionally, photovoltaic power generation has shown strong growth, with an installed capacity of 3 million kilowatts, and China's solar water heater collector area exceeds 200 million square meters. Further, China actively promotes the application of biogas, geothermal energy, tidal energy, and other forms of renewable energy. In 2011, the proportion of non-fossil fuels in relation to primary energy consumption was 8%, associated with an annual decrease of more than 600 million tons of CO_2 emission. It was anticipated that by the end of the 12th Five Year Plan, China's proportion of non-fossil energy consumption in relation to primary energy consumption would have reached 11.4%, and that the proportion of installed capacity for non-fossil energy power generation would reach 30%.

Rapid growth of forest resources greatly improves the carbon sequestration capacity and increases the potential for emission reduction. By 2020, China's forest cover is projected to reach 23%, and the forest stock volume will exceed 15 billion cubic meters. To achieve the target of a net increase of around 40 million hectares of forest area by 2020, compared with the forest area in 2005, the State Forestry Administration is stepping up efforts to organize and implement the National Afforestation Plan (2011–2020). In 2013, the afforested area, nationwide, was 91.50 million hectares, and the number of voluntarily planted trees was 2.52 billion, with both of these figures exceeding the annual plan targets. As of 2013, more than 20,000 hectares of carbon sink afforestation was completed in 18 provinces (autonomous regions and municipalities). China is actively promoting the transformation from the forest-tending subsidy pilots to the comprehensive forest operations management. It arranged RMB 5.8 billion of the central government's forest-tending subsidy to complete 118 million hectares of forest tending, exceeding the planned targets. Forests absorb a lot of CO_2, creating enormous ecological value for promoting the sustainable development of China, as well as for the global economy and society.

Over a longer historical duration, there are references to drastically reducing energy consumption in China. China's average annual GDP energy consumption fell to 5.2% during the period of the 6th Five Year Plan, rising slightly to 5.7% during the period of the 8th Five Year Plan, and up to 8.0% during the period of the 9th Five Year Plan. In 2013, China's energy consumption per RMB 1000 of GDP fell by 3.7% compared with its consumption during the previous year. During the first three years of the 12th Five Year Plan, the national energy consumption per unit of GDP dropped by 9.03%, enabling the achievement of the energy-saving target at about 350 million tons of standard coal. This was equivalent to a reduction of 840 million tons of CO_2 emissions, and it resulted in significant economic and social benefits. During the first half of 2014, the intensity of China's energy consumption was further reduced, and energy consumption per unit of GDP fell by 4.2%, which was the best performance since the beginning of the 12th Five Year Plan.[14] At the end of June 2014, the outcomes of optimizing the industrial structure conducted three times were 7.4%: 46.0%: 46.6%. Moreover, the proportion of industrial added value rose by 1.3% over the previous year, which exceeded that of secondary industry for more than six consecutive quarters. According to information from the National Leading Group to Address Climate Change and Energy Conservation and Pollutant Discharge Reduction, in 2014, China's energy consumption per unit of GDP and its CO_2 emissions decreased by 29.9% and 33.8%, respectively, compared with these amounts in 2005. Consequently, the emission reduction targets for the 12th Five Year Plan have been successfully completed. At a press conference held on July 27, 2015, representatives of the National Energy Administration revealed that during the period from 2000 to 2010, the average annual growth rate of China's energy consumption was 9.4%, while the average annual growth rate fell to 4.3% during the period from 2011 to 2014. It is anticipated that the growth rate during the period of the 13th Five Year Plan will further drop to around 3%. To this end, the core target of China's energy development plan during this current plan is achieving "four revolutions and one cooperation."[15] This is based on the premise of ensuring the energy supply, accelerating the development of non-fossil energy sources, improving energy efficiency, promoting energy-related institutional reforms and technological innovations, and promoting energy production and a consumption revolution.

To promote energy conservation and achieve a reduction in consumption, we first need to change the prevailing utilitarian approach, to be mentally prepared to fight a protracted war, and to commit to sustained basic work such as training specialized personnel, building an energy-saving information platform, and regularly evaluating energy-saving work. Introducing appropriate industrial policies can support a number of energy-saving and environmentally friendly manufacturers of technology and equipment, as well as service providers, and create new industrial growth points. Depending on conditions of national economic development and energy consumption, practical and feasible energy-saving targets can be developed, and adherence to energy saving as a long-term national policy can be ensured.

Whether government incentives and restrictions are effective mainly depends on their ability to change the process of investing in corporate resources and to guide enterprises to prioritize energy conservation over other investment options. The reduction of unit product energy consumption, along with other technical factors, are the direct causes of an increase in the energy consumption per RMB 10,000 of GDP. This indicates that cost-internalization measures taken by enterprises, such as improving energy efficiency through technological transformation and developing a circular economy,[16] actually increase their costs. Directly expanding businesses in terms of scale and the use of more energy are in fact more cost-effective. From this perspective, the problem of high energy consumption can be solved either by increasing the cost of business expansion, or by reducing costs for enterprises to undertake technological transformation and improve energy efficiency. At present, Chinese enterprises lack energy-saving system solutions, as well as energy-saving managers with the required expertise. These are basic issues that require urgent resolution. The government promotes energy-saving work through administrative means, but it is not a driving force for energy conservation. Because of the lack of basic energy data and energy managers, as well as of authorities with the expertise and ability to motivate them, enterprises are not motivated to reduce their emissions.

China still implements a low energy price policy, although the central government has demonstrated a very strong political will to promote energy conservation. Moreover, realistic and viable measures are lacking; for example, energy prices and energy-saving subsidies offset each other. Further, local governments are under considerable pressure because of the linkage between energy saving and governmental achievements.

China has successfully achieved the goal of reducing CO_2 emissions per unit of GDP by 17%, as proposed in the 12th Five Year Plan. The "U.S.-China Joint Announcement on Climate Change" and China's INDCs represent a solemn political commitment made by China to the world, and to its own people, to deal with climate change challenges and adapt to climate change, further demonstrating the Chinese government's political will and determination to tackle this key issue and develop a sustainable LCE through promoting energy conservation. Moreover, the political significance of China's first large-scale attempt to move toward an LCE is far greater than its economic significance.

On November 30, 2015, President Xi Jinping gave a speech entitled "Work Together to Build Win-Win Cooperation, Fair and Equitable Climate Change Governance Mechanisms" at the Climate Change Conference held in Paris. In his speech, he further reiterated China's target proposed in the INDC: namely, reaching a CO_2 emission peak by around 2030. He further stated that China would seek to achieve this goal as soon as possible. According to these targets, the amount of CO_2 emissions per unit of GDP in 2030 would drop by 60%–65% of the amount in 2005, while the forest stock volume would increase by about 4.5 billion cubic meters, and the proportion of non-fossil fuels in relation to primary energy consumption would reach about 20%. These targets are aligned to China's latest policies in response to climate change and express China's attitude and resolve regarding energy conservation. China will take its own practical environment-related steps to contribute to the global response to climate change.

6.6   Design of climate change policy targets and indicators in the 13th Five Year Plan (2016–2020)

In considering the design of the 13th Five Year Plan, **climate change has emerged as the most restrictive factor and the greatest constraint at home and abroad for China's development. Consequently, energy conservation and the response to climate change is now one of the core directives of national development policies. Along with the innovation in this plan, this is the greatest difference in comparison with previous Five Year Plans or planning processes.**

The core goal of China's policy is to build a climate-adapted society to address global climate change.[17] The natural environment is the foundation for building a moderately prosperous society by 2020. Resource conservation, environmental protection, ecological security, and disaster prevention and reduction are not only China's core national interests, but are also constitutive of China's long-term basic national policy in the 21st century. This includes the following five dimensions.

(1) **Constructing a resource-saving society** by adhering to a policy that combines development and conservation and considers preservation as the first priority. At the core of this policy is improving efficient use of energy and resources, with a focus on saving energy, water, land, and materials, and on developing resource-saving production, transportation, and consumption patterns.

(2) **Constructing an environment-friendly society** by preventing and controlling pollution at the source, reducing emissions of pollutants, and improving the quality of people's lives and the production environment. Other aspects include respect for the laws of nature, strengthening conservation, promoting ecological restoration, building areas differentiated by key functions, and reversing the trend of ecological deterioration.

(3) **Developing a circular economy** by promoting resource recycling and clean production, reducing polluting emissions at the source, and improving the cyclic ratio and ecological level of productive activities.

(4) **Developing an LCE** by exploring economic models entailing low energy consumption, low emissions, and low GHG emissions. An LCE requires developing low-carbon products, technologies, and energy, as well as innovative low-carbon consumption patterns. Developing green energy, especially renewable energy, further increasing forest cover and stock volumes, and forest carbon sequestration are also important aspects. Lastly, establishing a domestic carbon-emission trading system and implementing green taxes and a green trading system are components of an LCE.

(5) **Implementing national integrated disaster prevention and reduction strategies** by comprehensively enhancing the ability of the country, and of the entire society, to resist natural disasters and effectively safeguard people's lives and property is a final key policy dimension.

The stance taken in the 13th Five Year Plan is to vigorously promote a green revolution, build an upgraded green economy, improve the overall quality of the ecological environment, and achieve sustainable economic and social development.

China's 12th Five Year Plan included eight direct and three indirect green development targets. The direct targets have included decreasing energy consumption per unit of GDP by 20%; increasing the proportion of renewable energy to 14% (newly added index); increasing the proportion of clean use of coal by 10% (newly added index); decreasing SO_2 emissions by 10%; decreasing chemical oxygen demand emissions by 10%; decreasing CO_2 emissions by 10%, or decreasing CO_2 emissions per unit of GDP by 20% (newly added index); increasing the forest cover rate by 1.5%–2.0%; and reducing the proportion of direct economic losses from natural disasters to GDP to less than 1.5% (newly added index). The above have been defined as binding targets.

Indirect targets have included increasing the proportion of added value of the service industry in relation to GDP by at least 3%; increasing the proportion of R&D expenditure in relation to GDP by at least 0.3%; and increasing the urbanization rate by at least 3%. The above have been defined as expected targets.

In 2014, China's energy consumption per unit of GDP fell by 4.8%, and during the first half of 2015, it fell by another 5.9%. It was predicted that by the end of 2015, which fell within the period of China's 12th Five Year Plan, energy consumption per unit of GDP and carbon emissions per unit of GDP would cumulatively decrease by 18% and 19%, respectively, outperforming the targets of a reduction in energy consumption (16%) and carbon intensity (17%) prescribed in the 12th Five Year Plan. Moreover, according to the predictions made by different institutions, during the period of the 13th Five Year Plan (2016–2020), China only needs to decrease energy consumption per unit of GDP by 13% and carbon emission per unit of GDP by 14%–15% to meet its international commitments relating to emission reduction. These entail a 40%–45% decrease in energy consumption per unit of GDP by 2020 compared with energy consumption in 2005, and a decrease in carbon emissions per unit of GDP by 60%–65% by 2030 compared with the amount in 2005.[18] Developing a policy framework in accordance with China's conditions, in line with the country's core interests and those relating to global human development, is required to address climate change. It is important for China not only to have an equal standing with other world powers, but to also strive to be at the forefront of the global green revolution, green energy, and green development movement. There is a need to study and develop China's national programs for addressing climate change during the period of 2016 to 2020, and to adopt key policy objectives and targets for 2020 as follows.

(1) Saving energy: reducing energy consumption per unit of GDP by about 16%.
(2) Reducing emissions: reducing carbon emissions per unit of GDP by about 18%.[19]
(3) Innovating global green technology: establishing China's status as the 21st century's green energy revolution leader, innovator, and promoter.
(4) Developing the world's green energy market: China's provision of the world's largest market for wind and solar energy, its manufacture and export of new energy technologies and equipment, and its attainment of a 20% proportion of clean energy in relation to total energy in 2020.
(5) Expanding green space: China will provide the largest artificial forest sink, globally, and build the world's largest green ecological barrier ("Three-North Shelterbelt" and "Southeast Coastal Shelterbelt").

The above objectives have been designed with a focus on reducing GHG emissions and actively adapting to climate change. They are also the most important indicators for measuring and verifying whether China has adopted scientific development and achieved green growth that is measurable, evaluable, and comparable.

There are a number of important practices required for implementing the objectives/indexes of the 13th Five Year Plan. These include establishing an incentive-compatible green mechanism based on a top-level green development design; encouraging green technology innovation, setting the green benchmark for industries and implementing industry green standards; increasing green investments, creating a green finance platform, and promoting the development of green industry; establishing green accounting as a foundation and constantly improving the statistical carbon accounting system; improving the green legal system, strengthening regulatory measures and regulatory capacities, implementing a public information disclosure system and improving the effectiveness of green information dissemination; increasing ecological compensation funds, strengthening foundational activities for constructing an ecological compensation mechanism, and constantly improving prices, subsidies, and tax policies; and developing a green development performance index system entailing independent third-party assessment.

6.7 Highlights of the policy on climate change in the 13th Five Year Plan

The highlights in the 13th Five Year Plan for establishing various types of policies in response to climate change are presented below.

6.7.1 Extensive adjustment of the industrial structure

The plan places particular emphasis on reducing the proportion of industrial added value to GDP. The reasons for this are as follows. First, energy consumption per unit of industrial added value is the highest. In 2014, the proportion of industry accounted for 35.8% of the GDP, while the industrial power consumption accounted for 73.45% of total consumption, nationwide. Industrial energy consumption accounted for 62.7% of the total energy consumption, which was equivalent to 1.47 times the national energy consumption per unit of GDP, and 4.89 times the energy consumption per unit of national added value of the service industry.[20] Second, emissions of various pollutants per unit of industrial added value were the greatest. In 2014, industrial wastewater discharge accounted for 44.2% of the country's total volume, industrial COD emission accounted for 38% of the total volume, industrial SO_2 emissions accounted for 85.9% of the total volume, and industrial soot emissions accounted for 80.1% of the total volume in the country.[21]

6.7.2 An industrial development policy

Developing major energy-intensive industries (steel, building materials and non-metallic mining, and the chemical petrochemical industries) need to be strictly limited.[22] Moreover, accelerating the elimination of backward production capacity may enable China to avoid being locked into energy-intensive, carbon-intensive, and capital-intensive industries. China should mainly focus on developing high-tech industries that are intensive in relation to technology, human resources, and employment. Further, developing a modern service industry, especially a service industry that is intensive in relation to information and knowledge, as well as employment, should be prioritized.

6.7.3   Adjusting the energy policy

Adjusting the energy policy entails increasing the proportion of high-quality and renewable energy that is consumed and significantly limiting the application of a high-carbon energy structure. China's energy is considered the "dirtiest," globally. Moreover, its energy consumption structure is extremely irrational, being based on "dirty" coal, which has enormous social externalities and negative social costs.[23] During the period from 1996 to 2001, because of a decline in the domestic demand for energy, the government conducted a closed conversion to "clean" this coal for the first time. Consequently, the proportion of coal industry declined, but then showed a strong increase. It should be apparent that the thrust of future energy and industrial policies should be to strongly limit the coal industry through continued efforts to reduce the proportion of coal consumption, and to set main indexes to force effective policy enactment.[24] In addition, the national compulsory clean coal use ratio needs to be increased.[25]

6.7.4   Improved energy efficiency

Improving energy efficiency requires accomplishing the 2020 target provided in the Medium and Long-Term Energy Conservation Plan. Compulsory provisions on the average energy consumption target per unit of product output should be introduced every five years for energy-intensive industries, serving as the industry standard and creating market access barriers to motivate such enterprises to reduce their energy consumption.

Energy price reform policies should also be introduced, entailing a range of measures. These include fully liberalizing coal and electricity prices; implementing government-directed prices for electricity distribution and other naturally monopolistic businesses; promoting domestic oil and natural gas prices that approximate international prices and improving the pricing mechanism; and gradually liberalizing the wholesale and retail market sectors for petroleum, and encouraging market competition.

Introducing a pollution tax is crucial. Taxes on carbon, sulfur, and other pollutants should be levied. A carbon tax on imported and exported goods should be levied to internalize the external costs of coal and other "dirty" energies, and to force coal and power generation enterprises to adopt clean coal technology.

6.7.5   Science and technology policy

Technological innovation is the key to achieving China's emission reduction targets. Encouraging the introduction and use of all of the currently available major energy-saving technologies, worldwide, to reduce the costs of emission reduction would serve as the main medium- and short-term technical approach. The main medium- and long-term approach would be to encourage international cooperation in innovating green technology and protecting intellectual property. Developing various climate adaptation technologies, including agricultural, industrial, building, water-saving, and ecological environment protection technologies, is necessary, as is developing various climate adaptation technical standards.[26]

6.7.6   *Regional policy for emission reduction*

At the regional level, the HDIs of about 12 coastal provinces and cities (whose collective populations account for 40% of the total nationwide population) belong to the high HDI group, globally. During the period of the 13th Five Year Plan, these provinces and cities should implement the new binding target of achieving a 15% reduction in CO_2 emissions by 2020 compared with the level in 2015.

Optimizing the main targets of the development zones, according to "National Main Functional Area Planning," entails optimizing the industrial structure, reducing the proportion of heavy industry, and increasing the proportion of the service industry up to 73% in Beijing. The energy consumption structure should be optimized mainly through the use of domestic and foreign high-quality, clean, and new energy. This also requires achieving significant reductions in the proportion of coal consumption and CO_2 emissions, as well as partly disconnecting energy use from CO_2 emissions.

The collective population of the remaining 19 regions accounts for 60% of the country's total population, and their HDI levels fall within the upper-medium bracket, globally. Consequently, they should implement the conditional emission reduction targets during the 13th Five Year Plan.

6.7.7   *Expanding green space and developing forest-based industry*

Forests are the largest terrestrial sites of carbon storage and serve as the most cost-effective "carbon absorbers." According to one scientific calculation, the growth of trees per cubic meter can absorb 1.38 tons of CO_2, on average, and emit 1.62 tons of oxygen (O_2). Domestic experts in relevant fields have pointed out that the cost of storing one ton of CO_2 per hectare of forest is only RMB 122. Carbon storage within forests is both efficient and inexpensive, and it also creates more employment opportunities. China contains the largest artificial forest area in the world. It is estimated that from 1980 to 2005, the cumulative net absorption of CO_2 through China's afforestation program was about 3.06 billion tons. The cumulative net absorption of CO_2 through forest management was 1.62 billion tons, with a reduction of 430 million tons of CO_2 emitted through deforestation, which effectively enhanced the functionality of GHG sinks.[27]

6.7.8   *Implementing green investment and actively expanding the green new deal*

China, which is the world's largest clean energy investor, will launch a national carbon emission trading market in 2017. In 2014, the IEA predicted that over the next 50 years, natural gas will be the fastest-growing fossil fuel, with demand for this energy source growing by more than 50%. Increasing flexible global trade in liquefied natural gas reduces the risk of disruptions in gas supply. Moreover, China will become a key region promoting the growth of the global demand for natural gas. The proportion of China's coal demand to the global consumption of coal will stagnate at a level of a little over 50%, and will subsequently decline after 2030. By 2040, China, India, Indonesia, and Australia will account for 70% of global coal production. By 2040, China will account for 45% of global new nuclear power capacities, and by 2040, China's nuclear power will result in reducing the annual carbon emissions of countries with nuclear power of 8% (calculated as the percentage of predicted emissions at that time).[28] According to estimates made by the US Department of Energy, China's clean energy market was projected to reach US$186 billion by 2010, increasing up to US$555 billion by 2020.

Green trade should be supported and developed, and there is a further need to limit exports that are energy intensive and carbon intensive[29] and to cancel all kinds of disguised subsidies.

6.7.9   *International cooperation policies*

In recent years, extensive cooperation between China and the United States in the field of climate change mitigation demonstrates that it is possible for developed and developing countries to rise above their considerable differences and to jointly promote an international process of global cooperation to meet the challenge of climate change. This cooperation is not only indicative of a responsible power's attitude toward the world, but has also played a key role in the adoption of the Paris Agreement in 2015. In the future, China should be more actively involved in international cooperation on issues of new energy development, establishing international R&D funds, and technology transfer.

Notes

1 A low-carbon economy refers to the concept or form of economic development that seeks to maximize outputs based on the precondition of reducing GHG emissions. It is one of the basic strategies for combating human-induced climate change, as well as one of the core approaches associated with the new mode of global economic development in the 21st century. Even developed countries have only recently acquired a clear understanding of this problem. Theoretically, an LCE entails reduction of carbon content during both the production and consumption processes, cleaner air, and less CO_2 emissions. For example, renewable energy sources that are not associated with carbon emissions, low-carbon natural gas, and high-carbon coal all emit different amounts of carbon when consumed at the same level. The same coal that is consumed after being treated using clean coal technology will emit a much lower level of carbon than that emitted by the original high-carbon coal.

2 UK Energy White Paper, *Our Energy Future: Creating a Low Carbon Economy*, 2003. www.berr.gov.uk/energy/policy-strategy/energy-white-paper-2003/page21223.html.

3 Nicholas Stern, "Stern Review on the Economics of Climate Change," *NBER Working Paper 12741*. www.nber.org/papers/w12741.pdf.

4 Al Gore, "Gore's Challenge to the US: Make a Giant Leap for Humankind," *China Dialogue Website.* www.chinadialogue.net/article/show/single/ch/2274-Gore-s-challenge-to-the-US-make-a-giant-leap-for-humankind.
5 Hu Angang and Guang Qingyou, "Fighting Climate Change: China's Contribution," *Contemporary Asia-Pacific Studies* (2008).
6 See the interview of Reuters correspondent in 2008 Chris Buckley, with Professor Hu Angang. Chris Buckley, "China Government Adviser Urges Greenhouse Gas Cuts." www.reuters.com/article/reutersEdge/idUSPEK19898020080908.
7 Hanenergy Holding Group and China New Energy Chamber of Commerce released in China Conventional Center, Beijing, 2014.
8 Zheng Shuang, "How to Increase China's Competitiveness in World Carbon Market," *Energy of China* 5 (2008).
9 The World Bank, *State and Trends of the Carbon Market* (Washington, DC: The World Bank, 2008), 9.
10 These data were extracted from the China CDM website at http://cdm.ccchina.gov.cn/NewItemList.aspx.
11 Jiang Jiansi and Feng Chaoling, "Future of China's Low-Carbon Economy," *China Dialogue Website,* www.chinadialogue.net/article/show/single/ch/2330-Developing-China-s-low-carbon-economy.
12 Chen Xinhua, "Energy Conservation Needs Specific Theoretical Basis to Avoid Strategic Error," *Energy of China* 7 (2006): 7.
13 Project Team, Chatham House, *Changing Climate: Interdependencies on Energy and Climate Security for China and Europe* (Chatham House: Royal Institute of International Affairs in Britain, 2007), 12.
14 Data extracted from *China's Policies and Actions on Climate Change* (2014).
15 "Four Revolutions" are strategic ideas proposed by Xi Jim addressing China's energy revolution in 2014, which are energy consumption revolution, energy supply consumption, energy technology revolution, and energy institution revolution, respectively. "One Cooperation" means global energy cooperation to ensure energy security and win-winism.
16 Circular economy is a general term denoting reduction, reutilization, and resource recovery conducted during the processes of production, circulation, and consumption. Reduction refers to reducing resource consumption and waste production during production, circulation, consumption, and related processes. Reutilization refers to directly using waste as products, or after repairing, renovating and reproducing, or using all or part of the waste as a part of another product. Resource recovery refers to the direct use of waste as raw material or after recycling the waste. See the "Circular Economy Promotion Law of the People's Republic of China," which was examined and adopted during the 4th meeting of the Standing Committee of the 11th National People's Congress on August 29, 2008. This law was enacted by order of the president of the People's Republic of China, No. 4, and has been in force since January 1, 2009.
17 Eliminating greenhouse gases emitted into the atmosphere by humans will take a long time, and a global warming trend is inevitable. The fundamental way to address global climate change is through simultaneous emission reduction and adaptation. Constructing a climate-resilient society includes two key activities. The first is the design of a green development model that is suitable for the climate system and effective in reducing greenhouse gas emissions so as to slow global warming. The second is prospective adjustment of economic and social activities as an initiative toward adaptation to global warming.
18 See *21st Century Business Herald,* September 10, 2015. http://finance.sina.com.cn/chanjing/cyxw/20150910/005923203133.shtml?_t=t.
19 Jiang Kejuan, a researcher at the Energy Research Institute of the National Development and Reform Commission, holds that when developing emission reduction targets,

it is important not only to make the targets firmly binding (they should not be too low), but also to consider the problem of marginal abatement and the increasing effect of abatement costs. Consequently, it is appropriate to set energy consumption and carbon emissions per unit of GDP at about 16% and 18%, respectively, during the period of the 13th Five Year Plan. See *21st Century Business Herald,* September 10, 2015.
20 This is the author's estimate based on data for 2014 obtained from the National Bureau of Statistics.
21 National Bureau of Statistics and State Environmental Protection Administration, *China Environment Statistical Yearbook (2014)* (Beijing: China Statistics Press, 2015), 4–5.
22 IEA's statistics for 2007 indicate that a high energy consumption industry refers to an industry in which the proportion of energy consumption to total industrial consumption is more than 1.5 times the proportion of its industrial output in relation to the total industrial output. For example, in 2005 the industrial added value of China's three major sectors, namely, the steel industry, building materials and non-metallic mining industry, and the chemical and petrochemical industry, accounted for only one-fifth of the industrial added value, whereas energy consumption accounted for two-thirds of the total industrial energy consumption.
23 According to a quantitative estimation conducted by the Unirule Institute of Economics, based on coal prices and production levels in 2007, the direct external loss attributed to coal was about RMB 1.7903 trillion, which was equivalent to 7.3% of the GDP for that year. See *Report of Coal Cost, Price Formation and Internalization of External Costs,* March 27, 2009.
24 In 2005, the State Council introduced the goal of solving the problem of small coal mines within about three years and formulated, but did not complete, the "Notice of the State Administration of Work Safety on Amending Opinion Solicitation of Guiding Opinions on Strengthening the Administration of the Safety Base for Small Coal Mines." In 2008, there were 1,054 small coal mines, accounting for only 7.5% of the total number of coal mines. A total of 4.0 million tons/year of backward production capacity was eliminated, accounting for 1.6% of the total backward production capacity.
25 During the period from 2002 to 2007, the operational capacities of thermal power plants, nationwide, in terms of their flue-gas desulfurization mechanism, exceeded 270 million kilowatts, accounting for 50% of the thermal power capacity, and significantly reducing soot and SO_2 emissions. See Zhang Guobao, *China Energy Development Report 2009* (Beijing: Economic Science Press, 2009), 35.
26 Jiang Kejun lists eight major technologies: modern renewable energy technology (e.g., solar power); advanced nuclear power systems; fuel cells; an integrated gasification combined cycle, advanced clean coal technology, carbon capture, and carbon storage technologies; advanced gas turbines; unconventional natural gas and crude oil production technologies; synthetic fuel production technology; and ultra-low power consumption and zero-emission advanced transportation technologies. See Jiang Kejun, *China's Energy Demand and Greenhouse Gas Emission Scenarios;* and Yang Jiemian, *World's Climate Diplomacy and China's Response* (Beijing: Current Affairs Press, 2009).
27 A white paper on *China's Policies and Actions for Addressing Climate Change,* October 29, 2008.
28 International Energy Agency, *Executive Summary of World Energy Outlook 2014.*
29 In 2001, the proportion of China's net energy exports in relation to its total energy consumption was 18%, rising to 28% in 2004. CO_2 emissions amounted to about 1.1 billion tons, accounting for 23% of total emissions. See Huang Haifeng and Gao Nongnong, "Adjust the Industrial Structure and Open Up a New Path to Environmental Protection," *Environmental Protection* 12 (2009): 23.

7 Global significance and strategic consensus

Global climate change brings unprecedented opportunities and challenges to China. It also raises a number of questions. What responsibilities and obligations should China assume for tackling climate change? What basic ideas and policies should be proposed for this task? In the 21st century, China has entered a phase of accelerated industrialization. However, to tackle climate change, China will have to pursue green development, which can only be achieved using a new development mode and an alternative road to industrialization. China should actively reduce emissions to make a green contribution toward realizing the vision of "One World, One Dream, One Action" for our world. Therefore, this chapter proposes three major frameworks of global, national, and regional control to facilitate this process.

7.1 The focus of the debate on climate policy

China currently faces enormous international pressure[1] that emanates not only from developed countries, but also from less developed countries (for example, countries in Africa).[2] China also faces two diametrically opposed policy choices: **to implement or to disengage from global emission reduction.**[3]

China previously faced what can be described as a "role paradox" regarding the issue of global emission reduction. On the one hand, China is the world's greatest emitter of GHGs and bears the major responsibility for emission reduction. Without China's active participation, the efforts of developed countries cannot succeed, and all of the emission reduction programs implemented globally will ultimately fail. On the other hand, as a developing country, China does not want to take the lead in assuming its obligations for emission reduction, thereby offending other developing countries.[4] Therefore, China has not made any public commitment to reduce its emissions. Instead, it has strongly urged developed countries to reduce their emissions by 40%. However, through a process of consultation and compromise, which took place during international climate conferences, developed countries and developing countries have gradually reached a consensus that humanity is collectively facing the climate issue. Emission reduction is, therefore, a responsibility shared by all nations. Neither developed countries nor developing countries can independently bear responsibility for this. On

November 12, 2014, China and the United States made a joint announcement on climate change. China had previously declared quantified objectives relating to GHG emission reduction in 2009. However, in this joint announcement, China promised to curb its increase of CO_2 emissions before 2030. This amounts to making a promise regarding the emission peak, and reflects the assumption of responsibility expected of a great power.

Some Chinese scholars believe that global climate change is "their" (developed countries') problem rather than "our" problem (as a developing country). The author believes that global climate change should be viewed as a problem that concerns both "them" and "us," as well as China's own problem. In past disputes that occurred between developed and developing countries, each party contended that the problem of climate change concerned the other party, which should be held responsible for global climate change. However, the two parties have now reached a consensus that climate change is not just a problem that concerns either developed countries or developing countries. It is a problem that all of humanity is confronting, collectively. If we look, comparatively, at the climate situation of individual countries, China is not only the country that endures the gravest natural disasters; it is also facing the most serious air pollution problem affecting the largest population. The Chinese population, comprising more than one billion people, is the biggest victim of the climate change problem.

In the past, Chinese scholars widely believed that the reduction of global emissions was an action to be undertaken by "them" (developed countries) and not by "us" (developing countries). Moreover, had "we" participated in global emission reduction, this would have affected "our" development (black development). Currently developing countries have not yet undergone a process of industrial transformation, and highly polluting and energy-consuming industries remain prominent. Emission reduction will certainly affect the fast pace of their economic development. Consequently, they do not advocate emission reduction. The view presented in this book is that global emission reduction is an action that should not only be undertaken by developed countries, but also by developing countries, as well as specifically by China. One of the focal questions raised in the ongoing debate is how to understand "development." The perspective on development among Chinese leaders has changed from one that holds the speed of development to be the absolute principle to one that holds the quality of development to be the absolute principle. It has also shifted from faster development to scientific development. In recent years, the Chinese government has been paying increasing attention to climate problems. Chinese scholars and the public also demonstrate a positive attitude toward addressing climate change. For example, according to a report published by the BBC on the results of a worldwide survey, the perception of climate change is more widespread among the Chinese public compared with the extent of this perception among the publics of other Asian countries. Moreover, the Chinese public is more actively making efforts to respond to climate change.

China's domestic and international policies both address the climate issue. At the level of domestic policy, policymakers have reached a high degree of political consensus, guided by a scientific development outlook. China is the country that has achieved the best outcome relating to emission reduction and the most effective enforcement of emission reduction measures. However, at the level of international policy, views and opinions differ. The plan proposed at the Copenhagen Summit, held in December 2009, entailing mandatory emission reduction by developed countries and non-mandatory autonomous emission reduction by developing countries, has been demonstrated to be impracticable. Conscientious efforts made by major economies to meet their obligations of emission reduction constitute the most important actions that are required in the present. China's international policy on climate change is reflected in its submission to the UNFCCC, on June 30, 2015, of an INDC document titled *Strengthen Action Against Climate Change: Intended Nationally Determined Contributions of China,* which presented China's measures for tackling climate change. According to this document, by 2030, China will succeed in reducing CO_2 emissions per unit of GDP by 60%–65% compared with the amount in 2005. Attaining this goal will not only constitute the biggest challenge, entailing the greatest external pressure, but it will also offer the greatest opportunities and external conditions that will enable China to move forward together with other countries in the world, while assuming the role of an advocator, leader, and innovator of initiatives to reduce global emissions.

7.2 The domestic sphere and its significance: A shift toward green development

Addressing climate change and developing an LCE creates an opportunity for the realization of green development for China, which underwent industrialization at a relatively late stage. Consequently, there is no need at all for China to follow the same disastrous route taken by many Western countries, one that has entailed high resource consumption and significant pollutant emissions. Nor does China have to wait until it achieves high incomes for its population before implementing green development strategies. Instead, China can shorten the course of high resource consumption and high levels of pollutant emission, and embark on green development at a faster pace than evidenced for Western countries. However, environmental pollution and damage of the ecological environment, with the resulting adverse impacts on the air, soil, water, and other resources, have been prominent during recent years, creating an insurmountably wide gap between a beautiful and promising vision and the current reality. China needs to shift its mode of economic development from the conventionally implemented "black development" to "green development," turning away from ecological development and ecological deficit toward ecological construction and ecological surplus. However, China has resumed a mode of economic growth entailing black development characterized by high resource consumption and significant emissions of pollutants. Green development can only be realized if this developmental path is quickly replaced by a new path toward industrialization.[5]

In the 21st century, green development will rise to become China's new and innovative mode of development.[6] This mode conforms to China's

fundamental realities and is aligned with the country's national interests.[7] The 13th Five Year Plan clearly states that green development is a precondition for sustainable development and reflects people's pursuit of a better life. Therefore, China must adhere to core national policies of resource conservation and environmental protection, and to sustainable development and a civilized productive path that leads to a wealthy population and a healthy ecological environment. China must also speed up the pace of its efforts to build a resource-saving and environment-friendly society. It must develop new designs for modern construction, entailing harmonious development of people and nature, while promoting the construction of a beautiful environment and making new contributions to global ecological safety. General Secretary Xi Jinping has also emphasized that both environment and economic development are required and that the economy can never be developed at the expense of the ecological environment. China has publicly made this commitment to more than six billion people in the world.

The transformation of the economic development mode **is, in its essence, the transformation from a black to a green development mode**. This entails four correlative and complementary aspects as follows: (1) building a resource-saving society; (2) building an environment-friendly society; (3) strong advocacy for a recycling economy; and (4) strong advocacy for an LCE. The last aspect, in particular, represents a new mode of development. These dimensions reflect the close correlation and integration of China's green development with the world's low-carbon economic development.

At present, China is undergoing an accelerated phase of its industrialization process. During the period from 1979 to 2014, China's industrial added value grew at an annual average growth rate of 11.24%, increasing by 41 times. However, its industrialization process still characterized as black development, which is achieved at the cost of a high level of energy consumption and pollutant emission. In 2014, China's national industrial energy consumption amounted to 2.524 billion tons of standard oil, accounting for 62.78% of total end-use energy consumption. This figure is considerably higher than that for 1990 (36%) and far exceeds the average percentage (approximately 22%) among developed (OECD) countries.[8] This situation is closely linked to the prevalence of black industries across the country. At all spatial units – villages, towns, counties, cities, and provinces – industrial development is characterized by high energy consumption. This is inconsistent with China's national conditions and is contrary to the global development trend. Consequently, China's economic development is occurring at considerable costs relating to resources and the environment. Angang, Yuning, and Jinghai (2008) have studied China's green GDP following the initiation of the country's reform and opening up. They found that while natural capital losses have been diminishing, inputs of human capital have increased at a rapid rate. Further, China's ability to make use of global resources is expanding at a fast pace. Therefore, the average annual growth rate of "green GDP," defined by the World Bank for the period of 1978 to 2004, was 0.9% higher than the actual GDP growth rate. Angang et al. (2008) have argued that the annual average growth rate of green GDP should in fact be 1.22% higher than the actual GDP rate. The greater the loss of natural capital, the smaller the proportion of green GDP to actual GDP will be. The larger the input in human capita, the greater the proportion of green GDP to actual GDP will be. Enhanced abilities to use global resources (net imports of primary products) will result in an increase in the natural capital of the country and, consequently, in a higher proportion of green GDP. An increase in the growth rate of green GDP occurs in correlation with a reduction in the loss of natural capital, an increase in human capital inputs, and increased abilities to make use of global resources. Angang et al. (2008) have proposed a non-traditional mode of modern development. This contrasts with development modes adopted by Western industrialized countries, entailing support for and stimulation of high economic growth through a high level of resource consumption (especially non-renewable resources). The core underlying features of this alternative mode of modern development are as follows. The production system associated with this mode entails low resource consumption, complemented by a lifestyle characterized by moderate consumption. The associated economic system promotes stable economic growth and a constant increase of economic benefits. The social system provides guarantees for social benefits and equity. The technology system is characterized by constant innovation and complete absorption of new technologies, processes, and methods. Moreover, at the level of the international economic system, more open trading and non-trading promote closer connections with the global market. This modern development mode emphasizes rational resource development and exploitation, pollution prevention, and safeguarding of ecological balance. It features the following characteristics relating to resource and living consumption. During the first half of the 21st century, China's per capita consumption of all kinds of primary resources will generally be maintained at current levels, or will be slightly improved if it pursues this mode. The constraints imposed on consumption levels will require structural adjustment and quality improvement. In the long term, efforts should be made to maintain a relatively high level of accumulation and moderate consumption.[9]

It appears that China has been facing considerable external pressure, especially from Western countries, to reduce CO_2 emissions in the course of its future economic development. China should consider this problem rationally, objectively, and from a long-term perspective. Based on the country's basic conditions, its own long-term interests, and negative external global impacts, China should actively support global emission reduction initiatives, promote the formation of a new international mechanism, and avail of external constraint mechanisms to achieve a transformation of its mode of economic development. The decision made by the 15th Central Committee of the CPC regarding China's WTO accession was to refrain from surrendering to Western countries. This has resulted in promising opportunities for China to participate in international cooperation and exchange initiatives and to improve China's international competitiveness, thus providing a long-term opening to the outside world and the dividends of globalization.[10] Building on this successful experience, China should demonstrate decisiveness and determination regarding its participation in global efforts to reduce emissions. Because of international pressure to reduce global emissions, as well as the global opportunity to embark on low-carbon economic development, China can achieve a number of outcomes. These include optimizing its domestic industrial structure, restructuring traditional industries, developing industries with domestic characteristics, promoting the development of high technology and high value-added industries, and improving the proportion of the service industry in relation to the production and employment structures. Consequently, China will be able to achieve an LCE, as well as low carbon consumption.

Conventional economic practices are no longer appropriate. China must catch up with and join the current wave of low-carbon economic development through the accomplishment of a "great leap forward" in its development. This will enable it to establish new competitive advantages as soon as possible. Mohan Munasinghe, the vice chairperson of IPCC and chairperson of the Munasinghe Institute for Development in Sri Lanka, pointed out in a 2008 publication that developing countries should draw lessons from the experiences of countries that underwent industrialization before them. They should no longer follow their approach. Instead, they should take the route entailing a BDE curve that is lower than the BCE curve. Consequently, developing countries can find a sustainable development pathway that involves less pollution to reduce climate vulnerability. Threats and challenges to humanity caused by global climate change are becoming increasingly grave. During difficult times, special measures should be adopted. The current period is one such time, when global climate change is threatening human survival and economic development. China is one of the countries most affected by global climate change. In this case, the government undoubtedly provides the most powerful guarantee for advancing problem solving. In the general context of global warming, it is almost impossible for an LCE to be spontaneously formed. The government should forcefully promote an LCE with energy saving and consumption reduction to ensure that system change becomes mandatory and to eliminate market failure associated with external problems. The special measure for achieving this would require China's achievement of a one-step leap directly from a high-carbon to a low-carbon economy to guarantee energy and climate safety. This is the only option to enable China to succeed in effectively establishing a new comparative advantage in relation to industrial competition. Predictably, China will offer the world's largest markets in renewable energy, biological energy, clean coal, nuclear power, carbon trading, and environmental technology. In addition, it will have an LCE and will be an exporter of low-carbon products, as well as one of innovative low-carbon technologies in the future.

Assessed over a long duration, energy density and CO_2 intensity in key developed countries generally went through a process of first rising, followed by a period of decline, before maintaining a stable trend. In the 1980s, energy density in developed and developing countries were roughly equivalent. Subsequently, energy density in developed countries continued to decline while energy density in developing countries continued to rise. From the early 19th century onward, the CO_2 intensity of primary developed countries also exhibited a pattern of first rising and then declining. From the end of the 1940s, France, Britain, the United States, and Japan successively entered a stable period. In spite of the fact that developing countries reached their peaks in relation to CO_2 intensity much later than developed countries, they demonstrated essentially the same trend as developed countries. From the 1990s onward, the CO_2 intensities of China and India have tended to be stable at relatively low levels.

An assessment of China's long-term development trend indicates that the development of an LCE is the precondition for China to achieve peaceful ascendance in the world. For a large developing country like China, participation in global governance, drawing on the experiences of developed countries, articulation of China's perspectives during international negotiations, and promoting and realizing good domestic governance through global governance rules that are more widely applicable, are critical. These actions will enable China to achieve integration into the global system and international market, assume international responsibility and obligations, and increase its influence internationally. Based on China's experience of accession into WTO, the country's participation in the global system of governance and international rules could also become a catalyst for domestic reforms, as long as appropriate countermeasures are taken.[11] China's participation in international negotiations and global governance on climate change, and its acceptance of some or all of the current climate regulations, can provide the impetus and opportunity for China to realize good governance in relation to its energy and environmental policies.

China's response to climate change and its participation in international climate negotiations marks the first occasion of its proactive participation in the development of major international regulations. Currently, the responses of the political leaders of various countries to climate change reveal a consensus. The focus of the debate, now, is not *whether* climate change exists, but *how* we should respond to it. Therefore, China's focus now should be not on *whether* to participate in international climate negotiations, but on *how to obtain the right to speak* and become one of the leaders in the formulation of new global rules. Participating in international negotiations on how to respond to climate change and develop LCEs, as well as in the formulation of international rules, is of great significance. As previously mentioned, this marks the first occasion of China's active participation, in light of its new identity and status as a key global player, in negotiations and the development of international rules to regulate the new mode of global development.

7.3   Global significance: Making green contributions to human development

As pointed out by a number of scholars, current challenges relating to energy, the environment, and climate change, faced by various countries, cannot be solved under the existing global energy governance framework. International governance in the global energy field can be traced all the way back to the establishment of the Coal and Steel Common Market.[12] In fact, OPEC was the first global organization established to coordinate energy politics. The purpose of establishing an international energy organization was to coordinate global energy policies among member countries. OPEC's member countries are mainly members of economic cooperation and development organizations. At present, OPEC and IEA are the key international energy organizations in the world. In addition, there are some mechanisms for discussing energy problems such as the G8 Summit, regular global dialogue mechanisms between energy-consuming and energy-producing countries, and a number of institutional mechanisms, including the United Nations Non-governmental organizations such as the World Energy Council also play an important role in the field of international energy.[13] In general, however, China has not achieved a significant degree of participation in global energy governance. China's status in global energy governance is far from commensurate with its status as the second-largest energy consumer in the world. We strongly agree with the consensus reached at the 2014 United Nations Climate Change Conference, held in Lima, that all countries should endeavor to fight climate change together. What we, as a global community, lack is not natural resources or capital, but time. Some policies aimed at easing energy insecurity could also mitigate localized air pollution and climate change. Energy-consuming and energy-producing countries should earnestly discuss modalities of comprehensive cooperation to deal with challenges relating to energy, the environment, and climate change. A new energy governance framework should be established to deal with such challenges.

In the course of researching the energy policies of various countries, we found that energy-producing and energy-consuming countries, as well as important international organizations, have already realized that in the face of energy challenges, crises can be avoided only through cooperation. Because of significant differences among stakeholders regarding their interests, energy-producing and energy-consuming countries hold contrasting attitudes toward energy challenges. International organizations seeking "encompassing interests" also find significant discrepancies between various stakeholders. Non-cooperation between stakeholders poses the greatest risk within the global energy market. The greatest corresponding difficulty entailed in global energy governance is facilitating coordinated and collective action undertaken by various countries. At the core of global energy governance is the need to discover the encompassing interests of diverse stakeholders to enable collective action to be achieved and to develop and implement practical, feasible, and effective energy and environmental policies.

Discussions on global energy and climate governance should address the following five key topics of contention. The first topic for discussion is the inclusion of emerging high-energy consumers like China and India in the global energy governance framework. China is the world's second-largest energy consumer. However, as it is not a member of OECD, it cannot join the IEA. Efforts are required to promote the cooperation of China and India with the IEA. Effecting changes within the IEC would enable these two countries to become members of the IEA. Alternatively, in the absence of such changes within the IEA, consideration should be given to China's participation in the IEA emergency share system to promote information sharing and joint action. On this basis, certain international obligations can be assumed. This requires China's and India's inclusion in the entire international energy cooperation system. Without the participation of China, the IEA cannot fully actualize its key functions. Global energy governance can hardly be realized through the exclusion of the second-largest energy-consuming country from the key global energy organization.

The second key topic concerns the sharing of innovations in energy technology and improving energy use efficiency. It should be noted that on the one hand, China and India have newly emerged, globally, as large energy consumers. They can promote increased investments in energy-producing countries and propel the growth in energy supplies. Additionally, they are big markets for equipment and technologies relating to energy saving, emission reduction, and responses to climate change, as well as for alternative and renewable energy sources. Thus, an important global opportunity exists in relation to major energy-consuming countries. In 2006, China advocated a convention of the energy ministers from five countries: China, India, Japan, South Korea, and the United States. The "Joint Declaration by Energy Ministers of China, India, Japan, South Korea and the United States," published after the convention, indicated that some degree of consensus had been reached by the five countries regarding a diversified energy structure, energy saving, efficiency improvement, oil reserves, information sharing, business cooperation relating to energy, and other key considerations. In 2014, the leaders of China and the United States issued a joint announcement on the issue of climate change, announcing the two countries' action objectives for addressing climate change after 2020. We hope that energy-consuming countries will organize periodic discussions and cooperate with each other in such fields, collectively devising appropriate international mechanisms to cope with existing challenges.

The third discussion point entails guaranteeing a stable energy supply and demand and maintaining the stability of energy prices. This will require a variety of strategies. These include information sharing among energy producers and consumers, clear supply and demand signals delivered to the energy market to avoid sharp rises or crashes in prices, and discussions on the feasibility of adopting a variety of strong currencies as settlement currencies for the energy trade. Discussions on how to avoid sharp price rises of the world's resource and energy products, resulting from the devaluation tendency of the settlement currency, are also required. Energy consumers should offer energy producers advice and help in expanding their investments, strengthening technical development and cooperation, and expanding the energy supply through improvements in energy use efficiency and production technology.

The fourth discussion point concerns the need for coordinated efforts of the main economic entities on fiscal and monetary policies, reducing risks in the international financial market, and cracking down on excessive speculation in the oil market. To achieve the stability of the energy market first requires achieving the stability of the financial market in relation to energy. A sound mechanism needs to be devised for the international market, entailing the improvement of strategic petroleum reserves. Moreover, emergency preparedness and concerted reaction mechanisms are required for emergency response in various countries. Energy safety can only be guaranteed if the market is stable. Moreover, risks can only be dealt with collectively with the full cooperation of various countries.

The final topic of discussion concerns coping with the challenge of global warming and transforming global energy use. International cooperation requires joint adjustment of domestic energy policies by both energy-consuming and energy-producing countries, including China; reaching an international consensus; and taking consistent actions. Although the main energy-consuming and energy-producing countries in the world have taken certain measures relating to international cooperation, the global energy market still faces problems of sharp price fluctuations, tight supply and demand, and excessive GHG emissions. China should actively promote international climate talks, encouraging various countries to establish a global cooperation framework and a carbon emission market, along with widespread application of the CDM. China should also promote measures aimed at solving energy challenges that humanity collectively faces, such as those for strengthening international coordination.

Global energy governance should encompass not only cooperation between and among countries that are energy producers and consumers, and between these countries and international organizations, but also among international organizations. As a new player in the field of global energy governance, China should also be involved in the design of a new global energy governance framework. Consequently, it should contribute to good global energy governance and engage with energy, environment, and climate change challenges that face us all. China and the United States should strive to develop a new mode of cooperation, cultivating genuine cooperation with each other and promoting good governance worldwide. Effective communication between these two countries will enable major global issues to be identified and will provide the world with more public goods, thereby accomplishing the goals of global governance. China should consider becoming fully integrated into the international community and advocating for comprehensive inclusion as a stakeholder in global international organizations. China should also declare its intentions to join the G8 Group. Moreover, the United States should support China's accession to the IEA as one of the leaders in the domain of global energy governance, as well as a larger share of quotas and its right to have a say in the IMF. These efforts would promote the joint leadership of China and the United States in global responses to climate change.

China should make a greater contribution to human development on a global scale. Following its reform and opening up, China has contributed significantly to three dimensions of human development. These are global economic growth, global trade growth, and worldwide poverty reduction, which is its greatest development contribution.[14]

China's contributions in the abovementioned areas, especially poverty reduction, have drawn considerable attention and recognition worldwide. At the 18th National Congress of the Central Committee of the CPC, it was proposed that China must accelerate the transformation of its current mode of economic development. This was required to support the construction of an ecological civilization, to establish the basic industrial structure and modes of growth and consumption of energy, and to achieve effective conservation of resources, and of the ecological environment, by 2020.

In the coming decades, China should focus on two new aspects of its contribution to human development. These are its green and knowledge contributions. China's green contribution will be its most significant contribution to the world in the 21st century. Among the developing countries, China should take the lead in reducing its GHG emissions and achieve the record of encompassing the world's largest new forest-based carbon absorbing sink. China should publish details on its approach to emission reduction and attempt, by 2050, to reduce its carbon emissions to half of the amount in 1990.

In the short term, China's leaders should make full use of the window of opportunity provided by the Paris Climate Agreement to elaborate on China's emission reduction obligations. China should publish its roadmap to emission reduction, including a specific plan, at the earliest, and strengthen efforts to reach a global emission reduction agreement, thus demonstrating its status as one of leaders in the domain of global climate governance.

In the long term, China's leaders should consider low-carbon economic development and China's emission reduction obligations from the perspective of meeting the long-term interests of human development, as well as transforming its own economy and governance. This would constitute a truly green contribution to human development.

An LCE implies the way we choose to develop. Chinese culture advocates the integrity of nature and humanity. The significance of this integrity lies in knowing that, although resources and the environment are the basic preconditions for human survival, the development of human civilization has always been intertwined with nature. Having obtained adverse outcomes from the predatory exploitation and utilization of nature, evidenced in the development of our modern industrial civilization, we should rethink our existing ways of living, producing, and consuming collectively as the human race. We must acquire a real understanding of the material and natural world upon which we depend for our survival. As the great physicist Albert Einstein observed: "Holding the views we held when the problem occurred is useless for the solving of the problem."

The reason for our ability to survive and reproduce on the "Blue Planet" as the dominant species over millions of years lies in our capacity to rethink our behaviors and to cope consistently with all kinds of challenges that we encounter. Today, the turn to LCE development has brought us to a new crossroads. The greatest tests ahead relate to the choices we make and the decisions we take on our course of action.

7.4   Strategic consensus: Turning the challenge into an opportunity

Whether for the world or specifically for China, the global climate change phenomenon can be regarded as "unprecedented" in relation to two aspects: the "unprecedented" challenge it poses to development and the "unprecedented"

development opportunity that it creates. To turn this challenge into an opportunity requires unprecedented cooperation in place of confrontation. We also need to address common challenges collectively and to create a common development opportunity. Humanity will always develop, but will also face various challenges. Consequently, it has to deal with these challenges and, furthermore, to transform these challenges into opportunities. This transformation has been constitutive of the history of human development, and will continue to be constitutive of the future of human development. This study has led us to the following strategic consensus and basic conclusions.

(1) **Global warming poses an unprecedented challenge to the sustainable development of humanity.** As the IPCC's fifth report indicates, climate change will influence the basic elements of human life on a worldwide scale. These elements include water availability, food production, health, and the environment. Perceptions relating to the issue of climate change and warming, and its impacts across the world, have never appeared as heightened as they do now. This issue has thus emerged as the common challenge that humanity faces. While climate warming is a global phenomenon, its influences on different countries and regions vary, and the degree of damage resulting from the same influence may even vary among countries and regions at different stages of their development. In the same way, global climate warming poses an unprecedented challenge to China's sustainable development.

Let us recall China's basic conditions. China has a large population, a vast territory, diverse natural conditions, and a fragile ecological environment. Moreover, the country endures many natural disasters annually. According to the evidence provided in the IPCC's report, climate warming has caused the area of the Himalayan glaciers to shrink,[15] which will affect 250 million people in China to some extent, particularly those who are entirely dependent on fresh water originating from these glaciers. The low-level temperature rise in middle- and high-latitude areas of China can improve the growth conditions of crops by prolonging their growth or enabling new cultivated land to be developed. However, a further temperature rise will bring about lasting adverse effects such as more frequent increases to catastrophic temperature levels and intensification of shortages in water resources. Many major cities worldwide, including Shanghai, face the threat of flooding because of rising sea levels. In a country like China, which encompasses a vast territory, the climate may vary greatly from region to region. However, during the 20th century, the average ground air temperature of the whole country rose by 0.5–0.8 °C, with the temperature rise in northern China, and on the Tibetan Plateau, being more obvious than that in southern China.[16] The IPCC's fifth report suggests that the China's temperate and arid zones will extend northward as a result of the temperature rise. Cities such as Shanghai will experience more frequent and more severe heat waves that will adversely affect rapidly increasing urban populations. This projection reveals that China, as the most populous country, and covering a vast territory, is the principal victim of global climate warming. Based on China's own interests and sustainable development goals, the country is willing to adopt a more positive attitude and to take practical action in alignment with international action against global climate warming.

(2) **Responses to global climate warming constitute a typical example of GPGs, as well as an unprecedented opportunity for human beings to engage in sustainable development.** For many years, developed and developing countries have been opposing and confronting each other over a number of global development issues. However, when confronting the challenge of global climate warming, which affects all of humanity, they may join forces for the first time to work toward a global consensus and curb climate warming. For China, the issue of climate warming poses a challenge as well as an opportunity. We do not believe that there is a conflict of interest between China, as the world's largest developing country, and developed countries on this issue. On the contrary, China's interests are aligned with those of developed countries. Thus, to borrow from game theory, this is not a zero-sum game with a lose-win result; rather, it is a non-zero-sum game with mutual benefits and a win-win outcome. China's government is active on this issue. Moreover, China is the first developing country to undertake a country evaluation on climate warming and to propose goals, measures, and policies that are more or less consistent with those of the IPCC. The IPCC's report indicated a curbing of the global warming trend by 2015, as long as every country contributed 0.12% of its average GDP. A report released by the American Pew Charitable Trusts in April 2014 indicated that in 2013, China's investment in the field of clean energy accounted for 29% of the total investment in clean energy by the G20. Thus, China has overtaken the United States, now relegated to second place, and leads by US$17.5 billion. Furthermore, this is the third consecutive year that China has led the world. From this perspective, China can be considered to have the requisite capacity to take action in response to global climate warming, including investing a much higher proportion of capital than the amount recommended by the IPCC.

(3) **The historical debt and responsibility for global climate warming should be viewed and handled correctly.** Climate change is caused by the constant accumulation of GHGs in the atmosphere. These gases, emitted by human beings, for example, through car exhausts, remain in the atmosphere for a long time. CO_2 is the most common GHG, and it can remain in the atmosphere for up to a century. Therefore, the negative or negative externality "contributed" by a country to climate change is reflected not only in its current emission load (flow), but also in its accumulated emission load (stock) over a number of years. According to the estimation provided in the IPCC's Fifth Assessment report, commencing from 1850, the energy production and emissions of North America and Europe have collectively accounted for 70% of global CO_2 emissions. Of this amount, the United States accounts

for 30%, while the developing countries collectively account for less than 25%, with China's proportion being just 7%. Thus, the developed countries, and especially the United States, should bear historical responsibility in this regard. According to available data on GHG emissions released by Maplecroft (a world-renowned British company in the field of risk assessment), which included a ranking of countries with the highest CO_2 emissions in recent years,[17] the CO_2 emissions of five countries accounted for more than half of the total global CO_2 emissions (2012 data). The annual amount of GHG emissions reached 5.9 billion tons, and per capita CO_2 emissions amounted to 19.58 tons in the United States, ranking second after Australia. The data further revealed an expansion in Russia's large-scale industrial production from 1999 to 2005, resulting in a sharp increase in its annual CO_2 emissions to 1.7 billion tons and resulting in Russia being ranked third globally. The IPCC's report also noted that most future emissions would come from developing countries, associated with these countries' rapidly growing populations and GDPs. Moreover, the proportion of energy-intensive industries in these countries was found to be increasing. China emits more than 6 billion tons of CO_2 into the atmosphere every year and ranks first in the world for CO_2 emissions. However, China's per capita emissions are not the highest in the world. According to estimates, the accumulated emissions of developing countries is estimated to be equal to the emissions of developed countries by 2065.

(4) **Developed and developing countries should be simultaneously incorporated into the emission reduction schedule.** About 70%–80% of the existing CO_2 in the atmosphere was produced by developed countries. However, global climate change is "our" (common) problem, rather than being "your" (separate) problem. Whether emissions originate in New York or Shanghai, the effects of climate change are always the same for all people. Of course, those who are particularly vulnerable to climate change live in the poorest regions of the world.

In our opinion, because the carbon footprint of developed countries is the deepest, these countries should take the lead in clearly committing to their emission reduction obligations. However, developing countries should also be included in the schedule of countries that are committed to emission reduction obligations. Consequently, they should define the upper limit of their total emission reductions and their schedules for implementing their emission reduction obligations according to their national conditions and capabilities.[18]

The real difficulty is that the transition to an LCE still requires a number of coordinated policy and system interventions on an unprecedented scale. Therefore, providing financial and technical support is just one aspect of helping developing countries to respond to climate change. Helping these countries to consolidate their capabilities relating to systems and policies for responding to climate change is an urgent task. The global governance framework must provide developing countries with a set of combined practical and feasible policies and system references. This is because improved and independent capabilities among developed countries are required, based on their own economic transitions, for them to adapt to climate change.

Currently, global governance in relation to climate change responses has remained stuck in a prisoner's dilemma, with developed and developing countries not having reached a consensus. Without the active participation of developing countries, particularly large developing countries such as China and India, it is difficult to realize the goal of global emission reduction, and the situation may in fact worsen. According to the data presented in the Stern Report, without any actions being taken, GHG emissions will exceed 60 billion tons by 2023 and more than 85 billion tons by 2050.[19] Thus, the later emission reduction is implemented, the more radical the means and the greater the amount of emission reduction required. Consequently, it is imperative for countries across the world to set practical, feasible, and effective emission reduction goals and schedules. The populations of developing countries account for more than three-fourths of the global population, and the extent of damage that they suffer as a result of global climate change is much higher than that suffered by developed countries. The five rapidly developing countries, which include China and India, should face the reality of the serious consequences of climate change. In doing so, they should genuinely represent the long-term interests of developing countries, retract their position on not taking on the target of qualifying emission reductions, and actively participate in global efforts to reduce emissions.

(5) **The development of an LCE offers China an important opportunity to work toward and accomplish the transformation of the model of economic development** from a black one to a green one. The development of an LCE is helpful for increasing employment opportunities. During the process of developing a new LCE, China will make important investments and create job opportunities, as well as unlimited business opportunities. For example, about 7.7 million people in the world are currently engaged in the renewable energy industry, evidencing an increase by 18% compared with the number of people engaged in this industry (6.5 million) in 2014. On a global scale, the ability to create jobs in the renewable energy industry far exceeds the average industry level. In 2014, there were 2.5 million people engaged in the solar photovoltaic power generation industry, with China accounting for two-thirds of this workforce. While the number of jobs in this industry has also increased in Japan, it has been reduced in the EU countries. Other renewable energy industries also provide a substantial number of jobs, mainly relating to raw material supplies. These include the industries of biofuel (1.8 million employees), biomass energy (822,000 employees), and marsh gas (381,000 employees). The number of employees in the wind energy industry has exceeded one million people, half of whom are located in China. The number of people employed in the renewable energy industry in the United States, Brazil, and the EU countries is also on the rise. A total of 764,000 people are engaged in the solar hot water

and cooling industry, of whom three-fourths are in China, with other principal markets being India, Brazil, and the EU. There are 209,000 people engaged in the small-scale hydropower industry, half of whom are in China, with the remaining half being employed in EU countries, Brazil, and India. The large-scale hydropower industry directly provides approximately 1.5 million jobs, most of which are based in China and are mainly concentrated in the areas of construction and installation. A number of industrial and trade policies also exist, supporting employment in the renewable energy sector.[20] According to the *Renewable Energy and Jobs Annual Review 2015,* released by the International Renewable Energy Agency, there were more than 7.7 million employees, globally, in the renewable energy industry in 2015. Of these employees, approximately 3.4 million were engaged in the field of renewable energy in China, with the solar photovoltaic, liquid biofuel, and wind energy industries accounting for the most employees. There remains considerable potential for creating future job opportunities in the area of LCE development and protecting and cultivating forests.

(6) **China has the ability and conditions to deal with the challenge of climate change.** Appropriate inputs, including long-term inputs, are essential for effectively responding to climate change. However, the required quantity of such inputs is much smaller than is commonly thought. The 13th National Congress of the CPC estimated that the future growth rate of the Chinese economy would further slow at a rate of around 6.5%, and the annual average cost of emission reduction would have little influence on the development of the national economy. It is predicted that China's GDP will increase by 6.9 times between now and 2040 if adopting the LCE development mode. However, if not deploying the LCE mode, China's GDP will only increase by 7.2 times. Moreover, this estimate does not provide a complete portrayal of the huge gains for the Chinese economy brought about by emission reduction and the establishment of an LCE.

The pace of human development has been the most rapid over the last two centuries, evidencing a sustained process of industrial revolution and industrialization. What then is industrialization? In his book *Agriculture and Industrialization: A Preliminary Study on the Industrialization Problems in Agricultural Country,* published in 1949, Zhang Peigang provided a pioneering and theoretically foundational definition of industrialization. In Peigang's view, what initiated industrialization was an essential or strategic factor that could bring about a continuous change of the composition mode of a series of essential production functions in the national economy. In other words, this factor was the fundamental and determining force that could directly accelerate long-term sustainable economic growth, promote a change in socially productive forces, and facilitate the transformation of the social and economic structure.

The industrial revolution is not a phenomenon that occurs just once; rather, it occurs on a continual basis, constantly advancing from a low to an intermediate level, and subsequently from an intermediate to a high level. Beginning in 1750, there has been a succession of three worldwide industrial revolutions. During the first industrial revolution, the decisive factor for triggering the essential production function was the steam engine; during the second industrial revolution, this factor was electrical power and rail transport; and during the third industrial revolution, ICT products and technology constituted the key factor. However, the unfolding of these revolutions over a period of two centuries has been accompanied by an ever-widening gap between humans and nature. The outstanding performance of developed countries in relation to industrialization has entailed the emission of vast quantities of GHGs, accounting for approximately 85% of the historical cumulative amount of GHGs emitted since 1850. A fourth industrial revolution will be launched in the 21st century, namely the green industrial revolution. The revolution will accomplish the separation of the essential production function from carbon emissions and herald the development of green energy resources, industrial products, and consumption patterns.

China failed to participate in the first two industrial revolutions during the 18th and 19th centuries, and even during the first half of the 20th century, up to which time it remained closed and resistant to change. It was not until the reform and opening-up period that the country began to catch up with the third industrial revolution, namely the information revolution. In 1987, the penetration rate of telephones in China was only 2%. Consequently, informatization was absent and the country was essentially marginalized.[21] Because of its adherence to reform and its opening up to the outside world, China subsequently became a dynamic adopter of information technology. In 2000, the Chinese government clearly articulated a policy of implementing industrialization in conjunction with informatization,[22] and China embarked on a path toward becoming a potential leader in the area of informatization. China is currently the largest user, producer, and exporter of ICT technology and has become a leader in the area of informatization.

China's development model has had a huge impact on the world. Within its highly vulnerable ecological environment that accommodates the largest population in history, China's resource consumption and economic activity is also historically unprecedented. Consequently, the country is facing the most serious challenges that it has ever faced in relation to its ecological environment. These include large-scale and severe degradation of land resources, a deterioration of the hydrological environment, very severe degradation of grassland, and an expanded forest deficit; serious threats to biodiversity, continued high levels of urban air pollution, and an ongoing rise in the rates of natural disasters and hazards and economic loss. These impacts indicate that the processes of industrialization, urbanization, and modernization in China are also resulting in a rapid expansion of the gap between humans and nature. Restricted by its natural conditions, global resource supplies, and the environmental load, China is now being compelled to abandon the black development route. It can no longer follow the traditional path of industrialization, exemplified by the former Soviet Union. This entailed heavy industry with high energy consumption, high pollution, and low efficiency. Nor can it replicate and apply the modernization model adopted by developed countries, entailing high consumption and high emissions. Consequently, China must set forth on a new, green development road.

Upon entering the 21st century, the world stepped into a new stage of industrialization, namely the era of the green industrial revolution, which is essentially aimed at substantially improving resource productivity, reducing pollutant emissions, developing a circular economy and LCE (and even a non-carbonization economy), and disengaging economic growth and energy consumption from CO_2 emissions. We are at the dawn of the era of this new industrial revolution, and for the first time, China is playing a role as an initiator, a leader, and even an innovator, along with developed countries as well as other developing countries. Therefore, the fundamental direction of industrialization for China is new industrialization that serves to accelerate the transformation to green industry and promote an LCE, along with green energy resources and consumption and attainment of green development.

In November 1987, Deng Xiaoping proposed the strategic adoption of "three-step" modernization for China. During the third step of this process, China was expected to develop to the level of moderately developed countries by the middle of the 21st century – that is, by the 2050s. It was projected that China's per capita gross national product would quadruple, reaching a per capita GNP of US$4,000.[23] In 2007, China's per capita GDP had reached US$2,360. It is estimated that the per capita GDP will exceed US$4,000 by 2020, and that China will enter the global middle-income group, accomplishing Deng Xiaoping's third strategic step 30 years in advance of the projected goal. Currently, an important additional requirement is China's transformation into a green China by 2050. It is only through this transformation that complete modernization can genuinely be realized.

7.5   China's goal of "One World, One Dream, One Action"

Starting from the fundamental and common interests of China and of all of humanity, this study has sought to answer the following questions. What responsibilities and obligations should China bear in the response to global climate change? What key ideas and policy approaches should be proposed? Our basic argument throughout this study has been that **China should position itself in the vanguard of global emission reduction and make a green contribution to humankind.** Toward this end, we have proposed three governance frameworks at global, national, and local scales. Moreover, we have offered the following policy recommendations:

(1) **From the perspective of global governance, China should strengthen its role in international cooperation. It should participate in the new global governance framework that is currently dominated by the major developed countries and focus efforts on breaking the deadlock in negotiations on climate change.**

Because of the inequity in the sharing of costs and profits among various countries, international negotiations relating to climate change have reached a deadlock in the form of a prisoner's dilemma. Both for the world and for China, the time frame for acting is very tight, and we must make a choice within the shortest time possible. For every year that passes without an emission reduction agreement being reached, the GHG stock increases with a corresponding temperature rise in the future. No country is in a position to provide a GPG for addressing climate change on its own. It is impossible to achieve a governance framework that is accepted by all of the countries in the world using the UN-designed mechanism.[24] As an international convention, the Kyoto Protocol lacks the necessary punitive and incentive mechanisms. Moreover, only a few developed countries are subject to certain constraints, and these are only "soft constraints," based on a "soft index" and a "soft mechanism." Under realistic conditions, the major powers (large developed countries and large developing countries) should be proactive in taking actions, striving to reach a political consensus on an urgent basis, and consolidating and reinforcing the new global governance framework. On December 12, 2015, 196 contracting parties of the UNFCCC all agreed to ratify the new global climate agreement, known as the Paris Agreement, at the Climate Conference held in Paris. They further agreed to prepare for global actions to respond to climate change after 2020. All of the signatories to the Paris Agreement have declared that they will ensure that the rise in the global mean temperature does not exceed 2 °C before 2100, compared with the pre-industrial level, by supporting and reinforcing the global response to the threat of climate change. Further, they will strive to maintain the temperature rise within a limit of 1.5 °C. According to the Protocol, the signatories will declare their "intended contributions" relating to their participation in global actions that respond to climate change. Developed countries will continue to take the lead in reducing emissions and will simultaneously extend further support to developing countries to enhance their financial and technical abilities, thereby enabling them to take actions to mitigate and adapt to climate change. The developed countries have promised to allocate US$100 billion for this purpose each year, starting from 2020, increasing this amount before 2025. Consequently, China should support instead of oppose global governance. Moreover, it should participate actively rather than passively in global governance, committing itself to rather than escaping from its emission reduction obligations. China should actively contribute GPGs rather than opting for a free ride.

Global climate warming is a challenge that all of humanity must collectively face, and global governance is necessary for controlling and addressing this issue. We feel deeply that in the current context of economic globalization, the efforts of one country, on its own, are woefully inadequate for dealing with the global development issue. Global problems should be solved through intensive international cooperation in key areas that include political consultations, scientific research, technology, markets, and human resource development. From this perspective, China is willing to be part of global governance relating to climate warming.

Cooperation between China and the United States is not only necessary for both countries, but also for the world. China and the United States are the two greatest energy consumers globally, in addition to being the two greatest emitters of GHGs, accounting for more than 40% of total global emissions. In November

2014, these two countries issued a joint emission reduction statement. The United States planned to reduce its GHG emissions by 26%–28% from the level in 2005 by 2015. China promised to reach its emission peak by 2030 and to increase the proportion of non-fossil fuels to about 20%. At a difficult moment in international climate change negotiations, China and the United States, as the greatest emitters, succeeded in reaching an agreement, which was a "milestone" for global emission reductions. Long-term cooperation regarding energy and environmental protection between China and the United States has always been one of the main issues addressed in the China-America Strategic Economic Dialogue (SED). During the seventh round of the SED held in Washington on June 23–24, 2015, China and the United States declared their intention to make constructive efforts, together with other signatories of the UNFCCC, to ensure that green climate funds were effectively managed and utilized. Both countries were of the opinion that their bilateral investments in other countries, and the investments of international financial institutions and new development organizations, should support low-carbon technologies and climate endurance capacities. They further stated that they would consider shifting from the use of common resources for supporting and encouraging domestic and overseas projects to using low-carbon technology as a priority policy. Furthermore, both countries reaffirmed their commitment to specifying and gradually canceling inefficient fossil fuel subsidies that encourage waste. They promised to encourage other countries to participate in peer reviews on fossil fuel subsidies. SED is a platform and mechanism for fostering cooperation between China and the United States, enabling a frank exchange of views between the two countries on long-term and strategic economic issues. SED is playing an increasingly important role in relation to the economies of the two countries, and even in their bilateral relation. It fosters greater mutual and strategically based trust and understanding between China and the United States, helping to deepen cooperation in various fields. It is not only conducive to the fundamental interests of the two countries, but is also conducive to world peace, stability, and development. While SED is a bilateral mechanism, it will eventually develop into a multilateral dialogue mechanism. It is important that the SED platform is not weakened or abolished, but that it is enriched, reinforced, and developed, so that it can be used to further extend the space, fields, and issues of cooperation between China and the United States. Their cooperation in global issues, such as climate change, LCE, energy, and the environment, should be strengthened. The Sino-US bilateral "Ten-Year Energy and Environment" framework is the conduit for addressing the above issues. From now on, the two countries shall adequately communicate, exchange, and demonstrate mutual trust on these issues, promoting their gradual expansion from being bilateral issues to being multilateral and even global issues.

China and the United States should jointly promote the implementation of various deals under the Paris Agreement that address targets, mitigation, adaptation, loss and damage, capital, technology, capacity building, transparency, and a global inventory as soon as possible. Responding to climate change is the greatest challenge confronting human development in the 21st century. The participation of both China and the United States in actions aimed at addressing climate change in the form of "intended contributions" would constitute the material contribution required to achieve a reduction in global emissions. It would also mark the first comprehensive example of cooperation between China and the United States concerning major societal issues.

The Climate Conference held in Paris resulted in an agreement with lasting validity, is legally binding and responsible for future generations, reflects the most extensive consensus of the parties, and forms the critical node of global climate governance. As a responsible developing country, China should actively fulfill its relevant obligations and assume its international obligations that correspond to its overall conditions, development stage, and actual abilities. Moreover, it should continue to strive to achieve the action-oriented objective of responding to climate change before 2020, positively implement its "intended contributions," and strive to reach the peak as early as possible. Further, it should work together with other parties to support the implementation of the Paris Agreement and to expedite the establishment of a cooperative and win-win global climate governance system, in accordance with the principles of the Paris Convention.

(2) From the perspective of national governance, China should be clear about achieving its emission reduction schedule by 2050 and promulgating a national plan for the implementation of emission reductions.

China's emission reduction obligations should be considered rationally and objectively, based on the country's conditions, long-term interests, and negative externalities to the world caused by GHG emissions. Further, it should actively support and take the lead in the implementation of global emission reduction initiatives, promoting the establishment of a new international mechanism, which can help transform China's economic development model in turn. During the 18th session of the CPC Central Committee, based on China's own successful experience, it was proposed that a critical decision be taken, grasping the opportunity offered by global emission reductions and the development of an LSE. Other considerations identified by the committee included optimizing the domestic industrial structure, reorganizing and upgrading traditional industries, developing industries in line with national characteristics, promoting the development of high-tech and high value-added industries, improving the proportion of service industry within the production and employment structures, and achieving a national LCE, along with low-carbon consumption.

China's decision to take the lead in emission reduction is not a subjective one. Rather, it is determined by the special status and important role of China in the world. Regardless of whether or not China is willing, it must undertake its obligations to protect humanity and the living environment. It is a fact that China's per capita emissions are relatively low. However, this should not serve as an excuse to refuse to fulfill emission reduction obligations. China's significant influence on global climate change is indisputable, and if China denies this, China will be regarded as one of the obstacles in achieving global emission reduction targets, and even as the biggest obstacle. This would seriously damage China's international reputation and weaken its "soft" power.

China should, on its own initiative, actively participate in international climate negotiations and become one of the leaders and formulators of international rules. At present, a consensus has been reached by the leaders of all countries to respond to climate change. The current focus of the controversy is not *whether* there is climate change, but *how* to respond to climate change. Therefore, for China, the current focus, when considering this issue, is not *whether* to participate in international climate negotiations, but *how* to acquire the "right to speak" and become one of the leaders in formulating the new global rules. In the face of increasing international pressure and a growing conflict between resources and the environment, China's active participation in the formulation of new international rules could not only expand its own development space, but also promote the transformation of its model of economic development.

This study has focused on China's national governance in relation to climate change. In our opinion, China should actively participate in global carbon emission market transactions. It should simultaneously establish a domestic carbon emission trading market and set a pollution emission tax (carbon tax). The cost of China's emission reduction is comparatively low. From the international perspective, the Framework Convention stipulates that the cost per ton would be more than US$30. According to the calculation of Chinese scholars, the emission reduction cost in China would be only US$15 per ton. China has considerable potential regarding emission reduction based on a large market. Consideration can be given to setting a "carbon emission tax" in developed countries, as well as introducing this, more gradually, in developing countries, to enable the tax to be levied according to the carbon concentration emitted by producers and consumers. Such fees would be specifically used for conducting global scientific research, collecting and disseminating information, cultivating talent, conducting clean technology research, and engaging in development and afforestation.

China should gradually incorporate a resource tax, environmental tax, and fuel tax into its agenda. We believe that with the dawning of the era of the Paris Agreement, a carbon tax related to CO_2 should be inserted into the agenda to prompt China to take energy-saving and emission-reducing actions in response to climate change.

According to the 13th Five Year Plan for the National Economic and Social Development of the People's Republic of China, released by the Chinese government in 2015, the green concept will be the keynote of development during the next five years. The plan sets forth, for the first time, the objective of "generally improving the quality of the ecological environment" and proposes expediting the development of a system of property rights relating to natural resources, establishing and perfecting the property trading system and platform associated with the ecological environment, and initiating collection of an environmental protection tax. At the same time, it proposed to focus on promoting the transformation of the mode of energy production and use by concertedly promoting an energy revolution, optimizing the energy supply structure, improving the efficiency of energy use, constructing a clean, low-carbon, safe, and efficient modern energy system, and safeguarding national energy security.

According to the plan, compared with levels in 2015, by 2020, the water consumption per 10,000 Yuan GDP will decline by 23%, while unit GDP energy consumption will decline by 15%, the unit GDP CO_2 emissions will decline by 18%, and the share of non-fossil fuels in primary energy consumption will reach 15%. The forest cover rate will reach 23.04%, and the forest growing stock will expand to 16.5 billion cubic meters. The ratio of days of reasonably good air quality in cities at the prefectural level and above will extend to over 80%, and the concentration of fine particulate matter not reaching the acceptable standard in the cities at the prefectural level and above will drop by 18% compared with their concentration in 2015. The proportion of water at or exceeding Class III will be 70%, and the proportion of water below Class V will be less than 5%.[25] Chemical oxygen demand, ammonia nitrogen, sulfur dioxide, and nitrogen oxide emissions will respectively be 10%, 10%, 15%, and 15% less than those levels in 2015.[26]

(3) From the perspective of local governance, the central government should play a greater role in facilitating local governments and enterprises to respond to climate change.

If China makes a public commitment to the rest of the world regarding its emission reduction, and the central government sets a national emission reduction goal, China's domestic emission reduction schedule will consequently be accelerated. A study by Huanbo, Li, Huimin, and Ye (2008) revealed three major factors that caused local governments to respond to climate change. These were the central government's attention and call to respond, political mobility, and economic interest. The central government will play an important role in guiding the country's response to climate change, while regional and local governments should take further initiatives aimed at tackling climate change.[27] These efforts can provide the impetus and opportunity for China to implement good governance relating to its domestic energy and environmental policies through its participation in international negotiations and global governance relating to climate change, and its adoption of certain climate rules.

Therefore, local governments should be actively encouraged to reduce emissions according to the net carbon sources and HDI conditions of various Chinese provinces and municipalities. The effectiveness of the government's incentives and restrictive measures mainly depend on whether these measures lead to changes in enterprises' resource investment procedures through their prioritization of energy-saving investments and emission reduction above other investment options. At present, China's enterprises lack a systematic energy-saving and cost-reducing scheme and energy managers with professional energy-saving knowledge. These gaps require urgent attention. The government's promotion of energy-saving and cost-reducing work through administrative means has not provided the impetus for enterprises to save energy and reduce costs. Because of the shortage of basic energy data, energy managers, and authorization and incentives provided to relevant personnel, enterprises lack sufficient incentives for reducing emissions. Therefore, the compilation of basic statistical data should be strengthened and enterprises should be compelled to save energy and cut emissions through the implementation of energy price reforms.

With the fulfillment of the dream of attaining a moderately prosperous society based on 30 years of reform and opening up, China's vast population is in need of a new dream. This is provided by the concept of "One World, One Dream, and One Action."

"One World" essentially signifies that the world is getting smaller, and that countries are establishing closer relationships because of economic globalization. Therefore, to protect the earth is to protect our own home.

"One Dream" refers to a green dream. The earth we live in should be a green global village, and our world should be a green world. The world and China actually share this one dream. China's dream is to realize the world's dream.

"One Action" is the collective action of emission reduction implemented by humankind as a whole. This entails reducing GHGs by half by 2020.

As a country with a population of over one billion inhabitants, China should evidently assume responsibility by considering the dream of the world as its own dream and the actions of the world as its own actions. China should not only actively participate in the world's emission reduction, but should also play an exemplary role in making a green contribution to global efforts to reduce emissions.

Notes

1 According to an internationally widely held view, China would need to undertake massive emission reduction to realize the ultimate goal of "stabilizing the concentrations of greenhouse gases in the atmosphere at the level preventing the climate system from [the impact of] dangerous artificial interference" as specified in the Climate Change Convention. This will require massive emission reduction undertaken by China as a prerequisite. See Yang Jiemian, *Global Climate Change Diplomacy and China's Policy* (Beijing: Current Affairs Press, 2009), 265.

2 For example, the African Union, in its declaration adopted in Nairobi, observed that "China plays a very positive role [in] economic development in Africa, but China has become the world's largest emitter of greenhouse gases. Therefore, China must also take a lead at the Copenhagen Conference. As for Africa, the success of the Conference depends on whether the United States, China and Europe can . . . agree to sign the quota and trade agreement in light of seeking truth from facts and [based] on full understanding of the urgent situation and great efforts made by developing countries for mitigating and responding to climate change. . . . The United States, Europe and China [also need] to contribute to economic development in Africa and provide humanitarian aid." See Godwin Nnanna, "From Nairobi to Copenhagen," *China Dialogue* 7 (2009).

3 So far, the Chinese government has not made it clear whether it supports the goal of global emission reduction, which is to reduce global greenhouse gas emissions by half from their 1990 levels by 2050. It has also not clarified its national emission reduction objectives in relation to its industrial methods.

4 India, in particular, has refused to be a signatory to any binding international convention limiting CO_2 emissions.

5 Hu Angang and Guan Qingyou, "Four Feasibilities for China to Fight Global Climate Change," *Philosophy and Social Science Journal of Tsinghua University* 6 (2008).

6 This development mode entails the planning and construction of a green China. This encompasses a wide spectrum of activities, such as developing national ecological regional planning, improving social inputs in environmental protection, spending public finance on ecological assets, and engaging in ecological construction. The latter includes protecting natural forests, returning farmland and pasture to forests, regulating sandstorm sources, managing soil and water loss, wetland conservation, and desertification control for effectively curbing the deterioration of the ecological environment. These efforts will lead to an increase in the natural capital of the country. Increasing natural capital also entails innovating and developing green technologies, promoting green standards, developing a recycling economy, promoting clean production, using green energy, developing green industries, and eliminating technologies and production capacities entailing high energy consumption, high material consumption, and high pollution, as well as promoting and advocating green consumption, green foods, medicines, electrical appliances, furniture, cars, houses, buildings, and communities. This development mode further entails building green cities, designing and implementing green reforms, improving policy, evaluation, and legal systems, and developing compensation policies conducive to energy saving and environmental protection. At the levels of trade and product development, it entails conducting and promoting green trading; actively importing primary products, thereby increasing the country's natural capital; fully availing of new global technologies focusing on energy saving and environmental protection; and proactively developing products for export that meet international standards for environmental protection. At the level of international cooperation, key features include preventing the transfer of pollutants, actively carrying out international cooperation, observing international environment conventions, and proactively improving the global environment.

7 Hu Angang et al. posit that a country's fundamental interests and goals should relate to five major dimensions: national security and territorial integrity; economic development and economic stability; social justice and human security; political probity and social stability; and ecological balance and environmental protection. See Hu Angang et al., *The Second Transformation: The Construction of National Institutions* (Beijing: Tsinghua University Press, 2003).

8 International Energy Agency, *World Energy Outlook*.

9 Hu Angang, Gao Yuning, and Zheng Jinghai, "China's Green GDP and Green Productivity (1978–2005)," *Working Paper* (Beijing: The Institute for Contemporary China Studies Tsinghua University, 2008).

10 Hu Angang, "Primary Evaluation on China's Access to WTO: How China Will Affect the World Trade Growth Pattern (2000–2004)," *China Study* 6 (2006); Hu Angang and Men Honghua, "Re-Evaluation on China's Access to WTO: Full Opening, Full Participation, Full Cooperation and Full Enhancement of China," *China Study* 29 (2006).

11 China's accession to the WTO serves not only as a catalyst for domestic reforms, but also as one for foreign trade growth. Moreover, it provides a mechanism for "achievement sharing" among trading partners. Hu Angang, "Preliminary Evaluation on China's Accession to WTO: How China will affect the World Trade Growth Pattern (2000–2004)," *National State Report* (2006).

12 In 1950, the French foreign minister, Robert Schuman, proposed establishing the European Coal and Steel Community (ECSC) plan, seeking to entrust administrative authority for the French coal and steel sector to an independent body during the establishment of the coal and steel common market. Subsequently, six Western European countries, namely France, Germany, Italy, Belgium, the Netherlands, and Luxembourg, began negotiations on the basis of this plan. On April 18, 1951, these six countries signed the ECSC Treaty, entailing a validity period of 50 years in Paris. The ECSC was then formally established. This organization has contributed to the realization of the Schuman Plan and was subsequently further developed into the European Economic Community

(EEC). In 1967, its council of ministers and committees merged with corresponding organizations in the EEC.

13 Stainislav Z. Zhiznin, *International Energy Politics and Diplomacy,* trans. Qiang Xiaoyu et al. (Shanghai: East China Normal University Press, 2005), 47–49.

14 Beginning in 1820, the population of the absolute poor continued to rise, up to 1990. The main reason for the reversal of this trend after 1990 was the dramatic decline in the size of China's population. During the period from 1990 to 2005, China achieved a two-thirds reduction in its poverty. During the period from 1978 to 2005, the poor population in China was reduced from 250 million to 23.65 million. The poverty reduction goal of the UN Millennium Development Goals was to reduce the proportion of people whose average daily per capita expenditure was less than US$1 by half between 1990 and 2015, and to reduce the proportion of people suffering from hunger by half during this same period. China may be the first developing country to have accomplished the goal of reducing its extremely poor population (with an average daily per capita expenditure of less than US$1) by half, achieving a reduction from 32.9% in 1990 to 10% in 2005. Moreover, China has also made remarkable progress relating to hunger eradication, popularizing primary education, reducing the death rate of children, improving the health of pregnant women, and improving infrastructure and other social development indexes.

15 According to observations made by the Cold and Arid Regions Environmental and Engineering Research Institute of the Chinese Academy of Sciences, the eastern and western branches of the Tianshan Urumqi River Source No. 1 Glacier, which is one of the world's top 10 representative glaciers, are shrinking rapidly at a speed of 3.5 m and 5.9 m per year, respectively. This area has been reduced by 13.8% over a period of 45 years. July 26, 2007, China News Service, Lanzhou.

16 The research and analysis presented in *Climate in Tibet under the Global Warming Scenario,* published by the Climate Center of the Tibet Autonomous Region Meteorological Bureau, shows that the average temperature rise in Tibet was 7.5 times higher than the temperature rise at the national average level. The average annual temperature of Tibet is increasing at a rate of approximately 0.3 °C every decade. This increase rate is evidently higher than that of the national and global temperature increase rates. At present, China's temperature is increasing at the rate of 0.4 °C every century. Xinhua News Agency, July 23, 2007.

17 Maplecroft's "Ranking List of Global Top Ten Countries with Highest Greenhouse Gas Emissions."

18 The Blair Report indicates that to avoid the risks posed by an extreme climate, all countries should adjust the structures of their national economies and reduce their CO_2 emissions. The commitments of just the developed countries are not sufficient.

19 Nicholas Stern, "Stern Review on the Economics of Climate Change," *NBER Working Paper 12741.* www.nber.org/papers/w12741.pdf.

20 *The Annual Audit Report on Renewable Energy and Jobs of 2015,* newly released by the International Renewable Energy Agency.

21 UNDP, *Human Development Report,* 2001, Map 2.1.

22 During the 5th Plenary Session of the 15th Central Committee of the CPC, held in October 2000, it was proposed to accelerate informatization in relation to the national economy and society and to propel a combined process of industrialization and informatization. See Jiang Zemin, "Continuously Pushing Forward the Cause of Socialism Construction with Chinese Characteristics in the New Century," in *Selected Works of Jiang Zemin,* Volume III (Beijing: People's Publishing House, 2006), 127.

23 *A Chronicle of Deng Xiaoping,* compiled by the Party Literature Research Office of the CPC Central Committee (Beijing: Central Party Literature Press, 2004): 1, 217.

24 The Blair Report expressed concern about institutional capacities for responding to climate change. It noted that "although the United Nations Framework Convention

on Climate Change and Kyoto Protocol provided a good institutional foundation, they could not completely meet the requirements of effectively managing the response and action required for addressing climate change by them." See Tony Blair, "Breaking the Climate Deadlock: A Global Deal for Our Low-Carbon Future," *Report submitted to the G8 Hokkaido Toyako Summit* (2008) (Summary for Policymakers).

25 China has a set of standards in classifying the quality of its lakes and rivers, and there are five classes in total. Class III means the water quality is equal to that of the secondary protection zone of drinking water sources, general fish sanctuaries, and swimming areas.

26 Website of the Ministry of Environmental Protection of the People's Republic of China: http://zhoushengxian.mep.gov.cn/zhxx/hjyw/201603/t20160321_333874.htm.

27 Zhang Huanbo, Ma Li, Li Huimin, and Qi Ye, "Analysis of the Actions and Mechanism of China's Local Governments in Response to Climate Change," *Working Paper* (Beijing: School of Public Policy and Management, Tsinghua University, 2008).

Postscript

Global climate change is one of the greatest challenges ever to confront humanity, with the largest scale, the widest scope, and the most far-reaching influence, and it is also the largest constraint, the greatest challenge, and the greatest background affecting the economic and social development of China in future. As the world's largest developing country, China is an emerging industrialized country with the fastest-growing economy in the world facing the problem of how to industrialize rather than whether to engage in industrialization. Should we copy the Western model of industrialization? Or should we follow the model of industrialization of a traditional planned economy?

As a Chinese scholar, I have undergone a course of constant understanding of the study on the ecological environment and economic development in China. As early as 1989, we put forward a report on national conditions titled *Survival and Development* in the *China Study* series, arguing that China cannot follow the traditional modernization path of high living consumption, high resource consumption, and high pollution discharge (referring to per capita emission load) adopted by Western developed countries, but must take the new "non-traditional modernization path" of moderate living consumption, low resource consumption, and low pollution discharge that is different from that of the West.[1]

At the end of 1988, I, as the first author, completed the first report on national conditions titled *Survival and Development* for the National Condition Analysis and Research Team. The report put forward the development strategic vision of "ensuring the survival and sustainable development" for the first time, concisely generalizing the basic characteristics of China's national conditions and the basic national policies, adjudging that China would enter economic takeoff stage in around 1980 or so, analyzing the four plights of population, resource, environment, and food faced by the long-term development of China from the viewpoint of trend, explicitly presenting the argument that China must persist in fighting a "protracted battle" for realizing modernization, and proactively putting forward that China should choose a non-traditional modernization development model and adopt the strategy of "ensuring the survival and sustainable development." The core idea of this strategy is to implement a production system with low consumption of resources, a living system with moderate consumption, an economic system with continuous and steady economic growth and consistently improving economic efficiency, and an environmental protection system with reasonable resource developments and utilization, pollution prevention, and conservation of the ecological balance. Such an idea laid a solid foundation for the basic thinking based on which I carried out research on national conditions later, and it is also the earliest sprout of the idea of green development that I subsequently came up with in the later period.

In August 1989, Wang Yi, Niu Wenyuan, and I published the second report on the national conditions titled *The Ecological Deficit: Greatest Crisis of Survival of the Chinese Nation in Future* on behalf of the Ecological and Environmental Warning Research Team. The report analyzes that the background of China's "ecological deficit" is one of the seven worldwide ecological issues for the first time. The report considers the climate change and sea level rise as the first major challenge and believes that these changes will seriously affect China's more economically developed coastal regions and the entire agricultural production situation. The report also proactively puts forward that the ecological crisis is the biggest crisis, and also a common crisis, faced by humanity in the 21st century, which requires the joint action of all mankind. The report generalizes and assesses the situation of the ecological crisis of China, giving a warning to our whole society that we will face the most serious ecological crisis. We also specially point out in the report that China's rapid economic growth and the initial success of reform are at the price of the excessive consumption of resources and serious environmental damage. We not only grab too much and too fast from the already overexploited environmental resource account, but also consume the natural resources and environmental capital "borrowed" from our descendants, constantly expanding "ecological deficit." The report points out explicitly that "today's ecological environmental problem is closely interrelated with the social and economic activities of the entire world, and it will continuously evolve into a central issue of the survival and development of humanity in the 21st century. China not only will be severely affected by global climate change, but also will have a significant impact on the world's environment in turn." The report comes up with the basic idea of harmonious development based on "Nature – Economy – Society." This is the first time that we make the overall evaluation and trend analysis of China's ecological environment. The report attracted the attention of Comrade Song Jian, a then state councilor and director of the Environmental Protection Committee of the State Council, and he wrote down very touching written instructions.

On March 6–9, 1990, the Division of Earth Sciences of the Chinese Academy of Sciences convened a "Seminar on China Natural Disaster Analysis and Mitigation Measures" in Beijing. I, together with Huang Dingcheng and other researchers, wrote the *Proposals on Mitigation of Natural Disasters* (hereinafter referred to as *the Proposals*) on behalf of the Division of Earth Sciences of the Chinese Academy of Sciences, making a basic appraisal of the situation of natural disasters in China and concluding that China has a large population, complicated geological and geographical conditions, abnormal and changeable weather conditions, a fragile ecological environment, and various frequently occurring natural disasters, which cause great losses and damages. China is one of the few countries that are mostly affected by natural disasters. According to a preliminary estimate, natural disasters hit 300 million to 600 million *mu* of crops per annum in China, and the reduced grain yield caused by natural disasters reached more than 20 billion kilograms. In recent decades, the number of natural disasters increases with quickened frequency and more severe damage. The disaster area per annum throughout the country in the 1980s was 1.7 times the amount in the 1970s, and 2.1 times the amount in the 1950s. Natural disasters have seriously affected and restricted the sustained and stable development of the national economy, and it is also an important factor affecting social stability. We must pay close attention to the threats and ravaging of natural disasters faced by us, especially when we formulate national economic development plans and growth indicators, we should consider the constraints brought by natural disasters. In the *Proposals,* we also proposed strategies and measures for disaster mitigation, which include "strengthening scientific research, implementing disaster mitigation projects; further strengthening the government's leading role in disaster mitigation; including disaster mitigation work into the national economic plans practically, increasing financial investment in disaster mitigation; improving publicity and education, enhancing the people's awareness of disaster mitigation; and strengthening international cooperation."

In January 1997, we made a more detailed analysis of China's natural conditions in a book titled *Natural Disasters and Economic Development of China: From the Perspective of Physic-Geographical Environment* (written by Angang Hu, Zhongchen Lu, Wanying Sha, and Jianxin Yang, edited by Angang Hu, and published by Hubei Science Press in 1997): "China has a continental climate, significantly affected by monsoons with very wide scope, which is beneficial yet harmful to the agriculture, so China is [a] typical country with various types of natural disasters. From the perspective of history after the founding of the People's Republic of China, natural disasters in China have been aggravated constantly, and the number of middle-sized disasters amongst all natural disasters increases significantly with accelerated frequency of occurrence. The affected crop area keeps expanding and the disaster rate keeps rising. The crop area hit by disasters expands constantly with the continuously increasing impact of natural disasters on agriculture fluctuations, which has become an important constraint factor of China's economic development. We find that China is one of the countries affected most greatly by global climate change in the world. To this end, we put forward from a positive perspective that "disaster means the reduction of production, and the production reduction means the reduction of Gross National Product (GNP) or the cutting down of GNP; disaster mitigation means the increase of production, and the increase of production means the actual increase in GNP."

In 1998, an unprecedented flood occurred in China, which caused widespread concern at home and abroad. I presented several important conclusions in a report on the national conditions titled *Impact of Natural Disasters and Strategies for Disaster Reduction*: the amount of reduction in grain yield caused by disasters (flood, drought, frost, freezing temperature, hail and other disasters) of the country keeps rising every year. The proportion of loss to the total output increased from

2.1% in the 1950s to 5% in the 1990s. In the 1990s (1991–1997), the amount of average annual grain loss caused by natural disasters was 23.0287 million tons, equivalent to six times the amount of food losses in the 1950s, and about 5% of total production. Since the founding of New China, the economic loss caused by natural disasters grew, accounting for 3%–5% of the total GDP in the 1990s and 12%–40% of newly increased GDP in that year. Losses caused by various types of disasters throughout the country became more and more serious. The average affected population per annum in the 1990s accounted for one-third of the country's total population, and the average population of the disaster area per annum accounted for one-fifth of the country's total population. From the aspect of international comparison, China is one of the countries that suffer serious economic losses caused by natural disasters. At that time, when I was carrying out an analysis of the main causes of serious natural disasters, I pointed out that the first cause of the rare large-scale floods that occurred in China in 1998 was very unusual global climate change, of which the El Niño and La Niña phenomena were the "main culprit," and the second cause is the extremely complicated natural geographical environment in China. One of the root causes of floods caused by China's major rivers and lakes is the increasingly serious soil erosion, which has become China's number-one environmental problem. This flood also made me keenly aware of the need to "re-examine the relationship between man and nature."[2]

In recent years, I have put great effort into the research of China's response to the challenge of climate change. I have completed several articles and reports on national conditions, and I express my views by taking the opportunity to participate in domestic and international symposiums.

In August 2007, in the national condition report titled *How Does China Address the Challenges of Global Warming,* I put forward that

> Global Warming is an unprecedented challenge to sustainable development faced by mankind. As the world's most populous country with vast territory, China is [the] biggest victim of global warming . . . From the aspect of our own interests and the target of sustainable development, China is willing to adopt a more positive political attitude and more proactive practical actions in worldwide actions responding to global warming. In fact, without China's participation and action, actions taken by developed countries will not be successful. . . . On the positive side, it is an unprecedented opportunity for mankind to seek sustainable development, so we should correctly view and handle historical debts and historic responsibility of global climate warming.

My conclusion is that "China should strive to develop [a] low-carbon economy." China is a latecomer to industrialization, so it has both the opportunity to avoid the mistakes of other countries and the opportunity to innovate to create a new development model for the country. Therefore, we should avoid China's failure in environmental protection, so as to avoid the failure of the world; we should strive for China's success, so as to make the world achieve success in environmental protection.[3]

In November 2007, I was invited to comment on the report *World Energy Outlook 2007* in a release conference held by the International Energy Agency. I believe that the rise of China has made a great energy contribution to the world, has changed the energy pattern of the world, and has created a new pattern of global energy governance. Within the existing global energy governance framework, the problems of energy, environment, and climate change faced by various countries in the world cannot be resolved; the good governance of the world requires the participation of all countries and realizes collective action. Green development, green contribution, and green rise will be the greatest contributions made by China to the world. In the aspect of energy utilization, China should not play the role of a "follower," but should assume the responsibility of a "leader." It will be an important energy path for China's rise.

At the end of 2007, Guan Qingyou and I undertook the emergency research program titled "How Does China Addresses the Challenge of Global Climate Change," funded by the Center for Industrial Development and Environmental Governance (CIDEG) of the School of Public Policy and Management, Tsinghua University. We were beginning to carry out thematic studies on China's strategies and measures of addressing climate change.

In March 2008, the national condition report titled *Four Feasibilities for China to Fight Global Climate Changes,* we once again pointed out that the challenges of global climate changes faced by China can be divided into two major issues: (1) emission reduction and investment for climate change: costs and benefits; and (2) participation in international climate change negotiations: strategies and countermeasures. Both issues affect each other mutually. China should actively participate in international cooperation on climate change, actively participate in international climate negotiations, and make unequivocal commitments to emission reduction, so as to achieve sustainable development and green development of the domestic economy and society, assume the obligations of a responsible and big country for promoting the establishment of a harmonious world, and become a leader in addressing global climate change.

In July 2008, we translated *Breaking the Climate Deadlock: A Global Deal for a Low-Carbon Future,* a report prepared under the charge of former UK prime minister Tony Blair. The report was also "presented to the Group of Eight Summit in Hokkaido," with its political purpose of providing decision-making background to the G8 Summit held in July 2008 in Hokkaido, Japan. At that time, I attended a meeting with former UK prime minister Tony Blair titled "High-End Dialogue: The Tendency of Global Climate Change and China's Future Low-Carbon Road," held by the Chinese People's Institute of Foreign Affairs and the UK's Climate Group. I made a brief comment on the report, and the main points of the comments were sorted out and submitted to Tony Blair's office by the Climate Group. Afterwards, we specially compiled the national condition report *China Should Make a Public Commitment to International Emission Reduction Obligations and Implementation of Global Emission Reduction Timetable,* providing background support for China's leaders (former president Hu Jintao) to attend the G8 Summit. We explicitly pointed out that the costs and losses of emission reduction are limited, but the benefits are obvious, while the costs of no emission reduction are quite high. Both the developed and developing countries should adjust their economic structure to assume their emission reduction obligations. We maintain that China's strategic decision will have influence on the globe's well-being. It is a challenge for China's decision makers to seize the opportunity to gain initiative in tackling climate change. China has the ability and the need to actively participate in emission reduction, and China should make a public commitment to international emission reduction obligations and implementation of a global emission reduction timetable. China will become not only the biggest beneficiary of climate change, but also a leader and promoter in responding to climate change.

After the G8 Summit, I expressed my regret that China did not make any emission reduction commitments, and made my attitude clear on the climate change issue in a national condition report I wrote titled *China Should Stand at the Forefront of Emission Reduction and Make Green Contribution to the World.* This report indicated our clear position and attitude as a university-based think tank. Fortunately, however, the China government's position on this issues has been changing, and its policies are adjusting, too.

On October 29, 2008, the State Council Information Office published the white paper titled *China's Policies and Actions for Addressing Climate Change,* comprehensively introducing the effects of climate change on China, and published China's policies and actions to mitigate and adapt to climate changes before 2010, as well as the construction of institutional mechanisms made by China. However, this white paper did not publish the near-term, medium-term, and long-term goals and action plans to address climate change and emission reduction in China in 2015 (the ending period of the "12th Five Year Plan"), 2020 (building a moderately prosperous society comprehensively), and 2050 (the final target year of global emission reduction), for which I feel very regretful.

In November 2008, I put forward an ideal of "One World, One Dream, and One Action" in the national condition report titled *China Should Make Greater Green Contribution to the Human Development,* in hopes of pushing China to make changes on climate change issues.

Over the past year or more, I, as a research group expert, together with several other experts of the "Chinese Economists 50 Forum," took part in the special research on "Global Warming and China Economic Development," which was jointly carried out by the National Economic Research Institute of China Reform Foundation and the Swedish Environmental Research Institute, and we also participated in several discussions and international exchanges, presenting our views to international and domestic experts. As a consultant, I attended several meetings of the "China Low Carbon Economic Development Path" Research Group of the China Council for International Cooperation on Environment and Development. I presented my views to the experts attending the two conferences held in Stockholm and London, respectively, and received attention and praise from international experts. In addition, I also exchanged my views about our emission reduction program face-to-face with Lord Stern (Lord Nicholas Stern of the UK), Connie Hedegaard (Danish minister of climate and energy), and Susan Shirk (former US deputy assistant secretary of state and a China expert), as well as other experts, and gained wide support. Our global emission reduction program was praised as the most farsighted discussion[4] by Professor Kenneth Lieberthal, a famous American expert on China. In the UN climate meeting in Poznan, Poland, held in December 2008, Mexico committed to reduce emissions and became the first developing country that set a fixed amount of emission reductions with the goal of reducing its greenhouse gas emission level to that of 2002 by 2050, which undoubtedly strengthens the practical significance and operability of our emission reduction program.

At the end of 2008, we successfully completed the emergency research program funded by the Center for Industrial Development and Environmental Governance (CIDEG) of School of Public Policy and Management, Tsinghua University. On this basis, I wish to make further study on the near-term, medium-term, and long-term goals of emission reduction and action plans, as well as put forward package economic and social policies so as to form a "Tsinghua (a university-based think tank) Plan" for China's emission reduction, which will be the background research and policy recommendations for China to participate in the decision-making on emission reduction in future.

In the middle of July 2009, I was invited to visit the International Energy Agency in Paris, France to provide advice and comment on the *World Energy Outlook* of the current year issued by the International Energy Agency. After returning home, I once again reaffirmed the importance of China's commitment to emission reduction and the announcement of an emission reduction roadmap in the article titled *China's Climate Change Policy: Background, Objectives and Debate,* and made an initial design for China's climate policy goals during the period of the 12th Five Year Plan.

Then, entrusted by CIDEG, the Energy Foundation, and other agencies, we began carrying out a research project titled "China's Emission Reduction Goals and Action Plans." We hope that this research will further demonstrate China's emission reduction targets from the aspect of economic and technical feasibility, and we will carry out climate policy goal design more comprehensively and systematically. Since these targets will be used as background research and policy recommendations for China to participate in the decision-making on emission reduction and participate in international negotiations in future, it is imperative to study the countermeasures of China in the Climate Change Conference in Copenhagen in 2009, to propose the emission reduction targets of four stages: 2015, 2020, 2030, and 2050. I hope that our research can provide policymakers with timely "public knowledge" at such a critical moment, and I also hope that our research can influence public opinion on climate change issues.

Climate change has become a hot topic of academic research in China. This book is the result of many years of research by the Center for China Study, Tsinghua University. I took charge of the writing of this book. Dr. Guan Qinyou and I jointly wrote on this subject, and the report fully absorbed the views of experts in various fields. The two principles based on HDI emission reduction are one of our major innovations and are the result of our study of Chinese regional disparities.

As far as the judgment of the national condition concerned, we believe that the per capita income cannot demonstrate the comprehensive development capability of various places; therefore, HDI can more comprehensively reflect the level of development of a region. From a global point of view, we cannot delegate emission reduction obligations simply according to the degree of economic development, and such view can be said to be quite creative.

While studying this subject of China's response to climate change, we gained some experience. First, we need long-term understanding. Since there are quite big knowledge and information gaps in recognizing the relationship between man and nature, I recognized this problem through research of transition processes of natural disasters. I noticed the severity of China's natural disasters in the process of writing *Survival and Development*.[5] Second, we need international perspective. We should understand the impact of China on the world and the impact of the world on China, not only focusing on its positive impact, but also its negative effects. Third, we need independent policy recommendations. As a university-based think tank, it is necessary to put forward public policy objectives based on China's national interests and global interests and practical solutions separately. The establishment of national action plans will last until 2010, but we must study the situation of 2020, 2030, and even 2050 as a think tank, because climate change is an issue requiring long-term attention. Fourth, we should unequivocally express our views, not only dare to express our views in the internal policy discussions, and more importantly, we should also dare to express our views in international occasions, which is associated with certain risks. Today, decision-making has become increasingly democratized and specialized, and we have quite a lot of freedom to speak; therefore, we take the challenges of energy and environmental issues as our own challenges. This book only makes a preliminary study of this challenge, and we need to continue to study more about China's emissions reduction roadmap in 2020, 2030, and 2050.

This book was first finished in July 2009 and sent to China's decision makers for reading and reference. It was officially published in December 2009 by Tsinghua University. I understand for sure that the main points of the book are different from current domestic mainstream views. I expect that the publication of the book can trigger more discussion, providing more background knowledge and decision-making background for the Chinese government to participate in the United Nations Climate Change Conference in Copenhagen held at the end of 2009.

It is a shame that although China's leaders attended the Copenhagen climate conference, they did not make any public commitments of emission reduction after 2015, which disappointed the whole world. When conducting the research of the National 12th Five Year Program, we pointed out that China has become a target for the world's criticism in that it is now the biggest country in energy consumption and production, electricity consumption, sewage discharge, and CO_2 and SO_2 emission. Its transition in growth mode will experience heavier international pressure. Therefore, China must choose a green development mode and make contributions toward humanity's green development. We also proposed that during the period of the 12th Five Year Program, we need to implement policies to achieve the target of greatly reducing GHG emissions. We insist that by doing so, China's role and position in the world will also evolve correspondingly. In terms of global administrating (especially economic affairs), China will grow from followers to leaders. And in terms of global rules (especially those in tackling climate change), China will advance from passive accepters to rules-setting participants.[6]

On June 7, 2013, President Xi Jinping met President Obama at the Annenberg Retreat in California in the United States, and they both exchanged ideas on global climate change issues, laying foundations for future agreements between China and the United States.

On July 8, 2014, at the Experts Symposium on Economic Situation chaired by President Xi Jinping, I specifically proposed that there are four major constraining conditions that China's growth will be facing in future. The first is energy scarcity. The second is major scarcity of resources. The third is environment pollution. The fourth is climate change. I made a clear point that China is the country with the world's largest carbon emissions; moreover, its emission is equal to that of the US plus that of the EU, which will be the biggest constraining condition. To deal with these issues, China needs to maintain an appropriate economic growth rate, not the past's high-speed growth rate of 10%.

In November 2014, the heads of China and the US reached a consensus regarding this huge challenge of climate change facing all mankind and issued the "Sino-US Joint Statement on Climate Change," in which China and the US announced their action targets in responding to climate change after 2020. The US will achieve the targets of reducing 26% to 28% of emissions compared to that of 2005 and will strive for the target of 28%. China's object was to reach its emission peak by around 2030, and meanwhile its non-fossil energy's proportion in relation to primary energy consumption will reach around 20%. Both China and the US will put in more effort over time and jointly encourage the world to actively respond to climate change.

Later facts showed that a consensus on global climate change reached by both China and the US further facilitated the G20 political consensus and the UN General Assembly consensus, and thus the Paris Agreement was reached. The crucial opening year of fully implementing the Paris Agreement is 2016.

By now, President Xi Jinping and President Obama have issued three joint statements on climate change, which states clearly that China and America should sustain the momentum of exchange and cooperation to jointly push forward the enforcement of the Paris Agreement as soon as possible. We shall strengthen the confidence and resolution in participating in global climate governance from all levels, and will continue making positive contributions in building new Sino-US relationships and multilateral progress in climate change, which will be written into history as a global united response to climate change.

On September 3, 2016, President Xi and President Obama submitted the Paris Agreement ratified text to United Nations Secretary-General Ban Ki-moon. He stated that China and the US taking the lead in ratifying the Paris Agreement will greatly push it forward to come into force within this year. As a consequence, the proportion of carbon emission of countries participating in the Paris Agreement in relation to the globe will increase from 1.08% to 39.06%, with China's portion 20.09%, the United States' portion 17.89%, and China and the US in total 37.18%. The next step is that at least 55 contracting parties, their GHG emission accounting for 55% of the world's, will submit the ratified agreement and 30 days later the Paris Agreement will officially come into effect. It could be projected that after China and the US jointly ratify the Paris Agreement, it will officially come into effect soon and will start a new phase of global response to climate change.

These are also the objectives and expected achievements of writing this book in 2009. Though process is more important than results, process without any fruit is only process. From the publication of this book's Chinese version in 2009 to its English version in 2016, during these seven years we have already seen achievements.

Angang Hu, Qingyou, Guan

Tsinghua Park

August 30, 2016

Notes

1 National Condition Analysis and Research Group of Chinese Academy of Science: *Survival and Development,* Hu Angang, Wang Yi, Science Press, Beijing, 1989.
2 Hu Angang, "Disaster and Development: Impact of Natural Disasters and Strategies for Disaster Reduction in China," *China Study* 10 (1998).
3 Hu Angang, "How Does China Address the Challenges of Global Warming," *China Study* 29 (2007).
4 This is the most forward-leaning argument we have seen from a Chinese source.
5 When writing *Survival and Development,* I found that there is an ancient Chinese saying that "a famine occurs every three years, an epidemic occurs every six years, and a calamity occurs every twelve years," according to *Agricultural History of China* compiled by Mr. Tang Qiyu. During the past 2,200 years, major flood events broke out 1,600 times, and major drought events occurred 1,300 times, and often the floods and droughts occurred at the same time in different places. The number of disasters increases as time passes, with shorter time intervals. For example, the average number of disasters per annum is as follows: 0.6 times in the Sui Dynasty, 1.6 times in the Tang Dynasty, 1.8 times in the Song Dynasty, 3.2 times in the Yuan Dynasty, 3.7 times in the Ming Dynasty, and 3.8 times in the Qing Dynasty. In the last 30 years, the frequency of occurrence of drought in China's major arid and semiarid regions has quickened.
6 Hu Angang and Yan Yilong, "National 12th Five-Year Program: Background, Thinking Map and Targets," *National Research Report* 24 (August 8, 2010).

Bibliography

Chris Abbott, "An Uncertain Future: Law Enforcement, National Security and Climate Change," *Oxford Research Group*. http://oxfordresearchgroup.org.uk/sites/default/files/uncertainfuture.pdf.

Hu Angang, *China in 2020: Building Up a Well-Off Society*. Beijing: Tsinghua University Press, 2007.

Hu Angang, "Disaster and Development: Impact of Natural Disasters and Strategies for Disaster Reduction in China," *China Study* 10 (1998).

Hu Angang, "How Does China Addresses the Challenges of Global Warming," *China Study* 29 (2007).

Hu Angang, "Primary Evaluation on China's Access to WTO: How China Will Affect the World Trade Growth Pattern (2000–2004)," *China Study* 6 (2006).

Hu Angang and Men Honghua, "Re-Evaluation on China's Access to WTO: Full Opening, Full Participation, Full Cooperation and Full Enhancement of China," *China Study* 29 (2006).

Hu Angang and Zhang Ning, "Evolution of Regional Pattern and Disparities of China's Human Development (1982–2003)," in *National Report*. Beijing: Institute for Contemporary China Studies Tsinghua University, 2006.

Hu Angang and Guang Qingyou, "Fighting Climate Change: China's Contribution," *Contemporary Asia-Pacific Studies* (2008).

Hu Angang and Guang Qingyou, "Four Feasibilities for China to Fight Global Climate Change," *Philosophy and Social Science Journal of Tsinghua University* 6 (2008).

Hu Angang and Yan Yilong, "National 12th Five-Year Program: Background, Thinking Map and Targets," *National Research Report* 24 (August 8, 2010).

Hu Angang, Gao Yuning, and Zheng Jinghai, "China's Green GDP and Green Productivity (1978–2005)," *Working Paper*. Beijing: The Institute for Contemporary China Studies Tsinghua University, 2008.

Hu Angang et al., *The Second Transformation: The Construction of National Institutions*. Beijing: Tsinghua University Press, 2003.

Tom Titans Berg, *Environmental and Natural Resource Economics*. Beijing: Economic Science Press, 2003.

Tony Blair, "Breaking the Climate Deadlock: A Global Deal for Our Low-Carbon Future," *Report submitted to the G8 Hokkaido Toyako Summit*. 2008, http://blair.3cdn.net/b53ed18eb4812ef5d3_dem6be45a.pdf.

BP, Statistical Review 2013, *BP Energy Outlook 2035, Country and Regional Insights*. www.bp.com/en/global/corporate/about-bp/energy-economics/energy-outlook/country-and-regional-insights.html.

BP, *Statistical Review of World Energy*, 2014.

Joshua W. Busby, "Who Cares about the Weather? Climate Change and U.S. National Security," *Security Studies* 17 (2008): 468–504.

Kurt M. Campbell, Jay Gulledge, J. R. McNeill, et al., "The Age of Consequences: The Foreign Policy and National Security Implications of Global Climate Change," *Center for Strategic and International Studies*. www.sallan.org/pdf-docs/071105_ageofconsequences.pdf.

Chatham House (Royal Institute of International Affairs), *Interdependencies on Energy and Climate Security for China and Europe*, 2007.

China Ministry of Civil Affairs, *Social Service Development Statistical Bulletin 2010–2013* and the *Civil Affairs Development Statistical Bulletin 1999–2009*.

China National Bureau of Statistics, *China Statistical Yearbook*. Beijing: China Statistics Press.

China National Development and Reform Commission, *China's National Climate Change Programme*, 2007.

China.org.cn, "Fortune 500 Companies Still Increase Carbon Emission, Which Exceed the UN Emission Standards." http://news.china.com.cn/world/2014–12/25/content_34405638.htm.

China State Forestry Administration, *China Forestry and Ecological Construction Bulletin*, January 20, 2008.

Chronicle of Deng Xiaoping, Compiled by the Party Literature Research Office of the CPC Central Committee. Beijing: Central Party Literature Press, 2004.

William R. Cline, "Global Warming and Agriculture," *Finance and Development* 3 (2008).

The Coordination Group Office on Climate Change, the Management Center for Agendum in the 21st Century, *The Global Climate Change: Challenges Faced by Humanity*. Beijing: Commercial Press, 2004.

Feng Fei, "Accelerate Resource-Saving Society Construction and Promote Economic Growth Mode Transformation," *Electric Power Technologic Economics* 19 (3) (2007).

Al Gore, "Gore's Challenge to the US: Make a Giant Leap for Humankind," *China Dialogue Website*. www.chinadialogue.net/article/show/single/ch/2274-Gore-s-challenge-to-the-US-make-a-giant-leap-for-humankind.

Zhang Guobao, *China Energy Development Report 2009*. Beijing: Economic Science Press, 2009.

Zheng Guoguang, "Correctly Understand and Deal with Global Warming," *China Environment Observer* 1 (2007).

Xiao Guoliang, *Imperial Power and Chinese Society and Economy*. Beijing: Xinhua Publishing House, 1991.

Huang Haifeng and Gao Nongnong, "Adjust the Industrial Structure and Open Up a New Path to Environmental Protection," *Environmental Protection* 12 (2009).

The High Representative and the European Commission to the European Council, "Climate Change and International Security." www.consilium.europa.eu/ueDocs/cms_Data/docs/pressData/en/reports/99387.pdf.

Zhang Huanbo, "Climate Protection Policy Simulation Study: Macro Dynamic Economic Model Based on Multi-National Climate Protection." PhD diss., Institute of Policy and Management, Chinese Academy of Sciences, 2007.

Zhang Huanbo, Ma Li, Li Huimin, and Qi Ye, "Analysis of the Actions and Mechanism of China's Local Governments in Response to Climate Change," *Working Paper*. Beijing: School of Public Policy and Management, Tsinghua University, 2008.

Intergovernmental Panel on Climate Change, *Fifth Assessment Report*, 2014.

International Energy Agency, *2014 Key World Energy Statistics*.

International Energy Agency, *CO2 Emissions from Fuel Combustion Highlights*, 2014.

International Energy Agency, *Executive Summary of 2007 World Energy Outlook: China and India Insights* (in Chinese).

International Energy Agency, *The Impact of the Financial and Economic Crisis on Global Energy Investment*, 2009.

International Energy Agency, *World Energy Outlook 2007*.

International Energy Agency, *World Energy Outlook 2014*.

International Monetary Fund, *World Economic Outlook: Housing and the Business Cycle*, April 2008.

International Renewable Energy Agency, *The Annual Audit Report on Renewable Energy and Jobs of 2015*.

Interview of Reuters Correspondent in 2008, Chris Buckley, with Professor Hu Angang. Chris Buckley, "China Government Adviser Urges Greenhouse Gas Cuts." www.reuters.com/article/reutersEdge/idUSPEK19898020080908.

Jiang Jiansi and Feng Chaoling, "Future of China's Low-Carbon Economy," *China Dialogue Website*. www.chinadialogue.net/article/show/single/ch/2330-Developing-China-s-low-carbon-economy.

Yang Jiemian, *Global Climate Change Diplomacy and China's Policy*. Beijing: Current Affairs Press, 2009.

Cao Jing, Ling Jing, and Wang Li, "Possible Future International Climate Policy Model – Curbing Global Warming: 'Post-Kyoto Era' Where to Go," *Green Leaf* 5 (2008).

Ruan Junshi, *Ten Lessons on Meteorological Disasters*. Beijing: Meteorological Press, 2000.

Inge Kaul, Isabelle Grunberg, and Marc Stern, "Defining Global Public Goods," in *Global Public Goods: International Cooperation in the 21st Century*, eds. Inge Kaul, Isabelle Grunberg, and Marc Stern. New York: Oxford University Press, 1999.

Li Liping, Ren Yong, and Tian Chunxiu, "Analysis of China's Carbon Emission Responsibility from the International Trade Perspective," *Environmental Protection* 3 (2008).

Angus Maddison, *Chinese Economic Performance in the Long Run: 960–2030 AD*, trans. Wu Xiaoying and Ma Debin, proofread by Wang Xiaolu. Shanghai: Shanghai People's Publishing House, 2008.

McKinsey, *China's Green Revolution: Prioritizing Technologies to Achieve Energy and Environmental Sustainability*, 2009.

McKinsey, *A Cost Curve for Greenhouse Gas Reduction*, 2007.

Mohan Munasinghe, "Rising Temperatures, Rising Risks," *Finance & Development* 3 (2008).

The National Development and Reform Commission, *China's National Climate Change Programme*, 2007.

"The New Sputnik," *The New York Times*, September 27, 2009.

Godwin Nnanna, "From Nairobi to Copenhagen," *China Dialogue* 7 (2009).

Organization for Economic Cooperation and Development, *Governance in China (Chinese Version)*, trans. The Institute for Contemporary China Studies Tsinghua University. Beijing: Tsinghua University Press, 2007.

Population Census Office of the State Council & Department of Population Statistics, China National Bureau of Statistics, *1982 Population Census of China (Digital Collection)*. Beijing: China Statistics Press, 1985.

Eric Posner and Cass Sunstein, "Pay China to Cut Emissions," *The British Financial Times* (Chinese edition), August 9, 2007.

Project Team, Chatham House, *Changing Climate: Interdependencies on Energy and Climate Security for China and Europe*. Chatham House: Royal Institute of International Affairs in Britain, 2007.

Robert D. Putnam, "Diplomacy and Domestic Politics: The Logic of Two-Level Games," *International Organization* 42 (1988): 427–460.

Thomas Schelling, "Addressing Greenhouse Gases," in *Nobel Masters toward Energy and Environment-2007 Nobel Laureates Beijing Forum*, 178–179. Beijing: Science Press, 2008.

Zheng Shuang, "How to Increase China's Competitiveness in World Carbon Market," *Energy of China* 5 (2008).

Nicholas Stern, *Key Elements of a Global Deal on Climate Change*, 2008. www.lse.ac.uk/collections/granthamInstitute/publications/KeyElementsOfAGlobalDeal_30Apr08.pdf.

Nicholas Stern, *Revelation of Key Points of Global Climate Change Agreement on China*, transcript of a speech delivered at the School of Public Policy and Management, Tsinghua University, 2008 (not reviewed by the speaker).

Nicholas Stern, "Stern Review on the Economics of Climate Change," *NBER Working Paper 12741*. www.nber.org/papers/w12741.pdf.

Tsuneta Yano Commemoration, "Data Analysis of Japan 100 Years of Development," 2004.

Tsuneta Yano Commemoration, "Japan's Parliament," 2007.

UK Energy White Paper, *Our Energy Future: Creating a Low Carbon Economy*, 2003. www.berr.gov.uk/energy/policy-strategy/energy-white-paper-2003/page21223.html.

Unirule Institute of Economics, *Report of Coal Cost, Price Formation and Internalization of External Costs*, March 27, 2009.

United Nations, *Paris Agreement*, 2016.

United Nations Department of Economic and Social Affairs, *World Urbanization Prospects* (revised 2003 edition).

United Nations Development Programme, *China Human Development Report* (1997, 2002, 2012 and 2013), China Financial and Economic Publishing House.

United Nations Development Programme, *Human Development Report* (2001, 2013, 2014).

US Energy Information Administration, *Annual Energy Outlook 2015 with Projections to 2040*.

Tao Wang and Jim Watson, "Who Owns China's Carbon Emissions?" *Tyndall Briefing Note No. 23*, October 2007. http://tyndall.webapp1.uea.ac.uk/publications/briefing_notes/bn23.pdf.

Niu Wenyuan, *China Carbon Balance Trading Framework Research*. Beijing: Science Press, 2008.

World Bank, *2013 China Carbon Financial Outlook*.

World Bank, *State and Trends of the Carbon Market*. Washington, DC: The World Bank, 2008.

World Bank, *World Development Indicator 2006*, CD-ROM.

World Bank, *World Development Indicators 2007*, CD-ROM.

World Trade Organisation, *International Trade Statistics 2007*. www.wto.org.

Deng Xiaoping, *Chronicle of Deng Xiaoping (1975–1997)* (Part II), Compiled by the Party Literature Research Center of the Chinese Communist Party Central. Beijing: Literature Publishing House, 2004.

Deng Xiaoping, "Proceeding from the Reality of Primary Stage of Socialism," in *Selected Works of Deng Xiaoping*, Volume III. Beijing: People's Publishing House, 1993.

Chen Xinhua, "Energy Conservation Needs Specific Theoretical Basis to Avoid Strategic Error," *Energy of China* 7 (2006).

Matthew Yeomans, "Crude Politics: The United States, China, and the Race for Oil Security," *The Atlantic Monthly* 295 (4) (2005).

Wang Yi, "Exploring a Low-Carbon Development Path with Chinese Characteristics," *Green Leaf* 8 (2008).

Li Yu'e and Li Gao, "Status of Negotiations on the Climate Change Impacts and Adaptation," *Advances in Climate Change Research* 3 (5) (2007): 303–307.

Zhang Yuyan and Li Zenggang, *International Economic and Political Studies*. Shanghai: People's Publishing House, 2008.

Jiang Zemin, "Continuously Pushing Forward the Cause of Socialism Construction with Chinese Characteristics in the New Century," in *Selected Works of Jiang Zemin,* Volume III. Beijing: People's Publishing House, 2006.

Stainislav Z. Zhiznin, *International Energy Politics and Diplomacy,* trans. Qiang Xiaoyu et al. Shanghai: East China Normal University Press, 2005.

Index

Page numbers in italics indicate table or figure.

Abbott, Chris: *An Uncertain Future* 9
absolute poor 151n14
Action Plan on Energy Conservation, Emission Reduction and Low-Carbon Development 87
"Ad Hoc Working Group on the Durban Platform for Enhanced Action" 10
Alliance of Small Island States 12, 34
Allianz Insurance Group 35n1
Angang, Hu 129–30, 150n7; *The Ecological Deficit* 154; *Four Feasibilities for China to Fight Global Climate Change* 163; *How Does China Address the Challenges of Global Warming* 156, 157; *Impact of Natural Disasters and Strategies for Disaster Reduction* 155; *Natural Disasters and Economic Development of China* 155; *Proposals on Mitigation of Natural Disasters* 154
arid and humid areas *62*
Asia-Pacific Economic Cooperation 109
Australian Conservative Coalition 107

Bali Roadmap 10, 11, 33, 80
Blair, Tony: *Breaking the Climate Deadlock* (Blair Report) 26–8, 31, 32, 38, 39, 40, 90, 151n18, 151n24, 157; "High-End Dialogue" 157
BP: *World Energy Statistics 2013* 74
Busby, Joshua W.: "Who Cares about the Weather?" 8–9
Bush Administration 8, 13, 18, 52
Business as Usual Clean Development Mechanism *25*, 29

California Cap-and Trade Program 108
carbon dioxide (CO_2) 3, 9, 15–16, 22, *25*, 27, 28, 29, 30, *30*, *31*, 32, 35n10, 37, 39, 45, *48*, 50, 51, 65, *65*, 71, *72*, 74, 75, 76, 78, 85, 86, *86*, 87, 88, 90, 92–3, 94n1, 94n8, 96, 101, 103, 104, 105, 106, 109, 110, 113, 114, 115, 116, 117, 118, 122, 123n1, 125n29, 127, 128, 130, 131, 132, 138, 139, 143, 147, 148, 149n4, 151n18, 160
Carbon Dioxide Assessment Committee of the National Academy of Sciences 25
carbon footprint 31, *31*, 32, 35n10, 139
carbon trading 27, 40, 95, 101, *102*, 102–4, *103*
CCX *see* Chicago Climate Exchange
Center for a New American Security: *The Age of Consequences* 8
Center for Industrial Development and Environmental Governance 159; "How Does China Address the Challenge of Global Climate Change" 157
Center for Naval Analysis: *National Security and Climate Change Threat* 8
Central Intelligence Agency 9
CERs *see* Certified Emission Reductions
Certified Emission Reductions (CERs) 101, 104, *112*
Chaoling, Feng 112
Chatham Research Institute 75
Chicago Climate Exchange (CCX) 102, 102–3, 105, 106
China Agenda 21 86, 108
China and climate change 56–77; challenges posed by shift in global economic and trade patterns 75–7; global warming 61–4; natural disasters 56–61; resource and environment challenges for long-term development 64–74; *see also* national governance of climate change

China Human Development Report 2007/2008 5, 30, 34, 47, *83*, *84*, 86
China's Energy Policy (2012) 114
China Should Make a Public Commitment to International Emission Reduction Obligations and Implementation of Global Emission Reduction Timetable 157
China Should Make Greater Green Contribution to the Human Development 158
China's Policies and Actions for Addressing Climate Change 110, 158
Chinese Academy for Environmental Planning: *Coal Environment External Cost Accounting and Internalization Scheme Research* 74
Chinese Academy of Sciences 151n15, 154
circular economy 116, 118, 124n16, 143
"Circular Economy Promotion Law of the People's Republic of China" 124n16
Clean Development Mechanism (CDM) 10, 29, 101, 102, *103*, 104, 105, *105*, 111, *111*, *112*
Clean Power Plan 13
Climate Center of the Tibet Autonomous Region Meteorological Bureau: *Climate in Tibet under the Global Warming Scenario* 151n16
climate change negotiations 22–35; deadlock 13; emission reduction targets 22–6; international 16, 24, 30, 75, 145, 157; UN 10, 11
climate talks 10–17, 135
Cline, William 39
Cold and Arid Regions Environmental and Engineering Research Institute 151n15
common problem 30–3
Communist Party of China (CPC): National Congress 86, 91, 95, 108, 135, 141; *Report to the 17th National Congress of the Communist Party of China* 109; *Report to the 18th National Congress of the Communist Party of China* 109
Copenhagen Conference 1, 75, 149n2
CO2 *see* carbon dioxide
CPC *see* Communist Party of China

disasters, natural 56–61, *59*, *60*, *61*
Doha Conference 11
droughts 34, *57*, 58, *59*, 60, 162n5

earth's temperature 2
Ecofys 102; *G-8 Climate Scorecards 2008* 22, 35n1

ecological deficit 17, 128, 154
ecological disaster 4–6
economic disaster 6–7
Einstein, Albert 136
emission reduction, global 10–17, 19, *19*, 24, 25, 34, 37, 39, 46–54, 80, 91, 92, 101, 126, 127, 130, 136, 140, 143, 145, 146, 149n3, 158, 159; common problem 30–3; program that is acceptable to all countries 26–30; wealthy countries' carbon dioxide emissions in relation to global emissions *30*
Emission Reduction Fund 77, 107, 108
energy consumption 12, 20n15, 34, 56, *65*, 65–74, *67*, *68*, *73*, 76, 77, 79, 85, 86, *86*, 87, 88–89, *89*, 93, 98, 100, 109, 110, 118, 119, 120, 121, 122, 124n15, 124n19, 125n22, 125n29, 129, 142, 143, 148, 150n6, 160, 161; reducing per unit of GDP 112–17
Energy Foundation 74, 159
EU Cap-and Trade Program 103
EU Emissions Trading System (EU ETS) *102*, 102–4, *103*, 106
EU ETS *see* EU Emissions Trading System
European Coal and Steel Community 150n12
European Commission 103
Experts Symposium on Economic Situation 161

Fei, Feng 66–7
financial crisis 33–5, *42*, 50, 65, 98, 100; 2008 22, 51, 79
Five Year Plan 34, 94n2; 6th 115; 8th 115; 9th 85, 94n2, 115; 10th 67, 86; 11th 66, 86, 114; 12th 87, 92, 95, 102, 108, 110, 113, 114, 115, 116, 117, 118, 119, 158, 159; 13th 92, 116, 117–23, 124n19, 129, 147
floods and droughts *59*
"Four Revolutions" 116, 124n15
Framework Convention on Climate Change (UNFCCC), UN 1, 10, 11, 24, 26, 28, 29, 31, 33, 75, 104, 105, 108, 128, 144, 145, 147, 151n24; Paris Conference 75
Fukuda, Yasuo 27, 107

G8 22, 24, 28, *32*, 135; leaders 26, 27, 33, 36n11; Summit 24, 26, 27, 107, 133, 157, 158
G8 Climate Scorecards 2008 23, 35n1
G8+5 26, 32; Summit 10
GHGs *see* greenhouse gases

global emission reduction 10–17, 19,
19, 24, 25, 34, 37, 39, 80, 91, 92, 101,
126, 127, 130, 136, 140, 143, 145,
146, 149n3, 158, 159; classification
principles for reduction 46–9; common
problem 30–3; Obama's new energy
deal and prospects for reduction
49–54; program that is acceptable to
all countries 26–30; wealthy countries'
carbon dioxide emissions in relation to
global emissions *30*
global governance 37–54; best choice
for all countries 44–6; classification
principles for reduction 46–9;
establishing international mechanism
and collective action 41–4; losses
caused by global emissions 37–41;
Obama's new energy deal and prospects
for reduction 49–54
*Global New Energy Development Report
2014* 102
global public goods 18, *19*, *42*
global significance and strategic
consensus 126–49; domestic sphere
and its significance 128–32; focus of
debate on climate policy 126–8; global
significance 132–6; green contributions
to human development 132–6; green
development shift 128–32; "one world,
one dream, one action" 143–9; strategic
consensus 136–43
global warming 2, 3, 4, 6, 7, 9, 35n10,
42, 61–4, *79*, 88, 95, 124n17, 131, 135,
138, 156; sustainable development of
humanity 137; *see also* carbon footprint
Global Warming Prevention Headquarters 107
Gore, Al 97
Great Depression 7
green development 16–17, 87, 92, 93, 94,
97, 118, 119, 120, 124n17, 126, 128–32,
143, 154, 157, 160
green finance 33, 120
greenhouse effect 3, 20n5
greenhouse gases (GHGs) 3, 4, 6, 9, 12,
13, *14*, 18, 20n5, 24, *32*, 35n10, 44, 45,
72, *79*, 88, 90, 93, 94, 103, 104, 105,
106, 110, 124n17, 126, 138, 142, 144,
149, 149nn1–3, 159
Group of Eight Nations *see* G8
Grunberg, Isabelle 41
Guoguang, Zheng 20n3

Hardin, Garrett: *Tragedy of the Commons*
21n18

HDI *see* Human Development Index
Huanbo, Zhang 46, 148
Huang Dingcheng: *Proposals on
Mitigation of Natural Disasters* 154
human development 1–19; challenges
1–9; ecological disaster 4–6; economic
disaster 6–7; global emission reduction
10–17, 19; national security 7–9, 52,
150n7; political economy 17–19; social
disaster 7
Human Development Index (HDI) 46–9,
48, 55n16, 80, 81, *81*, 82, *83*, *84*, 92, 93,
94n7, 122, 148, 159, 160

IMF *see* International Monetary Fund
industrialization 12, 17, 56, 65, *66*, 67, 79,
86–7, 88, 92, 126, 128, 129, 131, 141,
142–3, 151n22, 153, 156
Intergovernmental Panel on Climate
Change (IPCC) 1, 7, 20n3, 43, 53, 61,
131, 139; Fifth Assessment Report 3,
4, *5*, 9, 11, 88, 95, 137, 138; Fourth
Assessment Report 10, 96; Working
Group III Report 37; Working Groups
55n9
international climate change negotiations
16, 24, 30, 75, 145, 157
International Energy Agency 10; *World
Energy Outlook* 159; *World Energy
Outlook 2007* 15, 32, 68, 71, 157
International Monetary Fund (IMF) 15, 39,
41, *42*, 43–4, 135
International Renewable Energy Agency:
Renewable Energy and Jobs 40, 141
IPCC *see* Intergovernmental Panel on
Climate Change

Japan Electric Power Exchange 107
Jianping, Huang 35
Jiansi, Jiang 112
Jing, Cao 28
Jing, Ling 28
Jinghai, Zheng 129
Jinping, Xi 161; "Work Together to
Build Win-Win Cooperation, Fair and
Equitable Climate Change Governance
Mechanisms" 117
Jintao, Hu 108

Kaul, Inge 41
Keidanren Voluntary Action Plan 106–7
Kejuan, Jiang 124n19
Kejun, Jiang 125n26
Ki-Moon, Ban 98, 161

Kyoto Protocol 10–1, 12, 13, 18, 22, *23*, 24, 26, 27, 28, 31, 44, 76, 80, 88, 100–8, 144, 151n24

LCE *see* low-carbon economy
Li, Wang 28
Lieberthal, Kenneth 159; *Overcoming Obstacles to U.S.-China Cooperation on Climate Change* 54
Little Ice Age 2
low-carbon economy (LCE) 17, 33, 34, 38, 39, 40, 56, 75, 76, 90–1, 92, 95–123, 123n1, 131, 156; bleak prospects for Kyoto Protocol and decentralization of global carbon market 100–8; energy-saving policy 108–12; essential for addressing the challenge of climate change 95–7; institutional energy-saving policy 110–1; mechanism of energy-saving policy 111; organizational energy-saving policy 111; reducing energy consumption per unit of GDP 112–17; strategic (national energy-saving policy) 109–10; 13th Five Year Plan 117–23; trend in global economic development 97–100
Lu, Zhongchen: *Natural Disasters and Economic Development of China* 155

McKinsey Global Institute: *China's Green Revolution* 40; *A Cost Curve for Greenhouse Gas Reduction* 40
Michaelowa, Axel 28
Montreal Convention 40
Montreal Protocol 55n14

Nasheed, Mohamed 6
National Afforestation Plan 115
National Climate Change Coordination Group 86
National Condition Analysis and Research Team 153
National Energy Administration 116
national governance of climate change 78–94; constraints compelling China to undertake emission reduction obligations 80–5; domestic context of China's emission reductions 85–91; economic development 1996–2006 85–6; emission reduction roadmap and green modernization 91–4; incorporation of factors during the 12th Five Year Plan 87; pressure from international community 87–91; prioritization of

mitigation 86–7; "Scramble" and "Blueprints" scenarios 78–80
National Intelligence Council 8
National Leading Group to Address Climate Change 109, 110, 111, 116
National New-Type Urbanization Plan 66
National Outline for Medium and Long Term S&T Development 76
National People's Congress 124n16
"National Plan for the Development of Science and Technology on Climate Change" 87
"National Plan on Climate Change of 2014–2020" 87
national security 7–9, 52, 150n7
National Strategy for Climate Change Adaptation 111
"National Strategy on Adaptation to Climate Change" 8
natural disasters 56–61
New South Wales Greenhouse Gas Abatement Scheme 103, 105
New York Mercantile Exchange 106

Obama, Barack 9, 13, 18, 49–53, 161
OECD *see* Organization for Economic Cooperation and Development
Official Development Assistance 49
"One Cooperation" 116, 124n15
"One Earth, Four Worlds," 47, *48*
"One World, One Dream, One Action" 126, 143–9, 158
Organization for Economic Cooperation and Development (OECD) 18, 20n15, 29, 31, *31*, 45, 66, *72*, 74, 129, 133
Our Energy Future 96
Outline for Promotion of Efforts to Prevent Global Warming 106
overconsumption 98

Paris Agreement 1, 11, 12, 24, 33, 35, 123, 136, 144, 145, 146, 147, 161–2
Paris Climate Conference 11, 22
Peigang, Zhang: *Agriculture and Industrialization* 141
Putnam, Robert D. 21n20

Qingyou, Guang 163: *Four Feasibilities for China to Fight Global Climate Change* 163
Qiyu, Tan: *Agricultural History of China* 162n5
Quebec Cap-and Trade Program 108

Regional Greenhouse Gas Initiative (RGGI) 105–6
research context 1–9; ecological disaster 4–6; economic disaster 6–7; national security 7–9, 52, 150n7; social disaster 7
Royal Dutch Shell: *Shell Energy Scenarios to 2050* 78

Sandalow, David: *Overcoming Obstacles to U.S.-China Cooperation on Climate Change* 54
Schelling, Thomas 25–6, 44
Schuman, Robert 150n12
sea level rise 4, *5*, *6*, 7, 154
"Seminar on China Natural Disaster Analysis and Mitigation Measures" 154
Sha, Wanying: *Natural Disasters and Economic Development of China* 155
"Sino-US Joint Statement on Climate Change" 33, 51, 161
snow 4
social disaster 7
State Council Information Office: *China's Policies and Actions for Addressing Climate Change* 158
State Forestry Administration 115
Stern, Marcus 41
Stern, Nicholas 6, 43, 44, 84, 158; *Key Elements of a Global Deal on Climate Change* 24; "Key Points to Address Climate Change" 10; Report 34, 140; "Stern Review on the Economics of Climate Change" 96
Strougal, Lubomir 94n10
Survival and Development 153, 160, 162n5

Tokyo Metropolitan Government 107
Tsinghua University 160; School of Public Policy and Management 43, 159

UN *see* United Nations
UNFCCC *see* United Nations: Framework Convention on Climate Change
Unirule Institute of Economics 125n23
United Nations (UN) *14*; Charter 97; Climate Change Conference 6, 10, 12, 33, 53, 117, 133, 159, 160; climate change negotiations 10, 11; Climate Summit 11; Environment and Development Conference 108; Framework Convention on Climate Change (UNFCCC) 1, 10, 11, 24, 26, 28, 29, 31, 33, 75, 104, 105, 108,

128, 144, 145, 147, 151n24; Millennium Development Goals 98, 151n14; Paris Agreement 1, 11, 12, 24, 33, 35, 123, 136, 144, 145, 146, 147, 161–2; Paris Conference 75; Security Council 17, *42*, 43
United Nations Development Programme 49, 61, 63, 104; *China Human Development Report 2007/2008* 5–6, 7, 30, 34, 44, 45, 46, 47, *83*, *84*, 86, 88
United Nations Environment Programme 43
United States Center for Strategic and International Studies: *The Age of Consequences* 8
United States Congress 18, 51
United States Department of Defense: *An Abrupt Climate Change Change Scenario and its Implications for United States National Security* 8
urbanization 65, 66–7, 79, 92, 119, 142
US Council on Foreign Relations: *Climate Change and National Security* 8
US Energy Information Administration (EIA) 52, 68

Warsaw Conference 11
wealthy countries' carbon dioxide emissions in relation to global emissions *30*
Wenyuan, Niu: *The Ecological Deficit* 154
WMO *see* World Meteorological Organization
Work Plan for Greenhouse Gas Emission Control during the 12th Five Year Plan Period 110
World Bank 6, 41, *42*, 43, 44, 84, 102, 105, 129
World Meteorological Organization (WMO) 43
World Trade Organization (WTO) 16, 41, *42*, 43–4, 97, 130, 132, 150n11
World Wildlife Fund 35n1
WTO see World Trade Organization

Xiaoping, Deng 94n6, 94n10, 143
Xinhua, Chen 113

Yang, Jianxin: *Natural Disasters and Economic Development of China* 155
Yi, Wang: *The Ecological Deficit* 154
Yuning, Gao 129

Milton Keynes UK
Ingram Content Group UK Ltd.
UKHW040054071024
449327UK00019B/556